The American Transportation Revolution

THE AMERICAN TRANSPORTATION REVOLUTION

A Social and Cultural History

Aaron W. Marrs

Johns Hopkins University Press
Baltimore

© 2024 Johns Hopkins University Press
All rights reserved. Published 2024
Printed in the United States of America on acid-free paper
2 4 6 8 9 7 5 3 1

Johns Hopkins University Press
2715 North Charles Street
Baltimore, Maryland 21218
www.press.jhu.edu

Library of Congress Cataloging-in-Publication Data

Names: Marrs, Aaron W., 1976– author.
Title: The American transportation revolution : a social
and cultural history / Aaron W. Marrs.
Description: Baltimore : Johns Hopkins University Press, [2024] |
Includes bibliographical references and index.
Identifiers: LCCN 2023030022 | ISBN 9781421448497 (hardcover) |
ISBN 9781421448503 (ebook)
Subjects: LCSH: Transportation—United States—History.
Classification: LCC HE203 .M287 2024 | DDC 388.0973—dc23/eng/20230629
LC record available at https://lccn.loc.gov/2023030022

A catalog record for this book is available from the British Library.

*Special discounts are available for bulk purchases of this book. For more
information, please contact Special Sales at specialsales@jh.edu.*

For Melissa Jane

CONTENTS

Acknowledgments ix

INTRODUCTION *1*

1 Community Relations *10*

2 Travel *38*

3 The Arts *72*

4 Religion *103*

5 Black Passengers *129*

6 White Women Passengers *165*

7 Children *187*

CONCLUSION *208*

Notes 211
Essay on Sources 249
Index 255

ACKNOWLEDGMENTS

It has taken me much longer than I anticipated to complete this book. I am grateful to Johns Hopkins University Press for keeping the faith through all the years, starting with Bob Brugger and finishing with Matt McAdam and Adriahna Conway, who ensured that I did not trip as I crossed the finish line.

It would have been impossible to do this research without the financial assistance that I received from the American Antiquarian Society, the Library Company of Philadelphia, and the Huntington Library. I want to highlight these three institutions in particular because they make their research grants open to scholars who do not have an academic affiliation. I urge other institutions to follow their example. I appreciate the support that Christopher Clark, John Lauritz Larson, Chandra Manning, and Amy Richter gave to my applications for these fellowships. My employer, the Office of the Historian at the US Department of State, permitted me to take a leave of absence to pursue these fellowships, for which I am thankful. I also used materials gathered on earlier research trips I took for my first book, which were sponsored by the Harvard Business School and the Virginia Historical Society. I thank all of those institutions in addition to the following libraries where I also conducted research: the Hargrett Library at the University of Georgia, the Historical Society of Pennsylvania, the Library of Congress, the Maryland Center for History and Culture, the Massachusetts Historical Society, the South Caroliniana Library at the University of South Carolina, and the Wisconsin Historical Society.

Portions of the book manuscript were presented at conferences. I am grateful to the audiences and commentators at the annual meetings of the Society for Historians of the Early American Republic (Christopher Clark), the Nineteenth-Century Studies Association (Elizabeth Boone), the Popular

Culture Association (Kathy Mason), and the Business History Conference (Rowena Olegario).

Two anonymous readers for Johns Hopkins University Press made helpful critiques. I also want to thank the following for reading some or all of the manuscript: Sara Berndt, Noralee Frankel, Chandra Manning, Rebecca Shrum, and Melissa Jane Taylor. Their incisive comments sharpened my focus and prevented numerous gaffes. The errors that remain are, of course, my own.

A wide range of family, friends, and colleagues (and one cat) have cheered me on from near and far and buoyed my spirits during the writing process. Finally, there is simply no way that any of this could have happened without the steadfast love and support of my wife, Melissa Jane Taylor. She believes in me even when I don't believe in myself. It's done, we did it, let's celebrate!

Introduction

Diarist Sallie McNeill lived with her family in Brazoria County, Texas, after graduating from Baylor University's Female Department in 1858. Her family had moved to that county in 1850. They came from Arkansas, hoping to find better-quality land in this region, which was newly annexed to the United States. When McNeill's family moved to Texas, there were no railroads in the state; railroad corporations had been chartered, but no construction had begun. By the decade's end, however, there had been some progress: a railroad reached into Brazoria County near McNeill's plantation home. In February 1860, just after turning twenty years old, McNeill saw a train for the first time. The event was significant enough that she recorded it in her diary: "I could hardly realize, that this was my first sight of the 'iron horse,' because I have read and heard of the cars so often, that everything seemed natural."[1]

Read and heard. The railroad, a complex technological achievement—by 1860 not new to the world but still new by sight to McNeill—had become so thoroughly integrated into American life that to McNeill it seemed completely "natural" on first glance. We do not know precisely what she had read or heard prior to that fateful day—a fictional story about a train ride in a magazine, spelling lessons featuring trains in her reader at school, grim jokes about accidents printed in a comic almanac, or perhaps all of these together. During the twenty years of McNeill's life, railroads and steamboats had had a large presence, not just on the physical landscape but on the cultural landscape as well.

In surveying this cultural landscape, we can see that antebellum Americans had abundant opportunities to encounter steam transit in their imaginations long before it arrived in their towns. Steam transit pervaded print and oral culture, including newspapers, sermons, poetry, humor, and fiction. Steam also entered the language as a metaphor, including on the most explosive political issues of the day, when enslaved people escaped to freedom via an "underground railroad." Images of railroads and steamboats could be found in cartoons and comic almanacs or carried in people's pockets on currency. Steam transit also changed the aural landscape: people could hear orchestras play songs that imitated railroads or were written in honor of steamboats. They could purchase the sheet music for similar songs to play on their own pianos at home. McNeill's exclamation of familiarity merely hints at the wide world of cultural expression that steam transit inhabited in the nineteenth century. Through all of these means—and many more—we can see how the machine became "natural" for Sallie McNeill.

Historians have long been attuned to the importance of transportation in early American life, but this has generally been in the context of political or economic history. Such work goes back decades to George Rogers Taylor and the concept of the "Transportation Revolution." The increased availability of steamboats (starting with Robert Fulton's invention in New York in 1807) and railroads (starting with construction in the late 1820s) unquestionably had a profound effect on American economic life. In the historiography of the antebellum United States, scholars have generally treated transportation, or "internal improvements," as a condition for whatever larger economic transformations they wished to discuss. In the closing decades of the twentieth century, those arguments centered around the "market revolution"—its timing, spread, and impact.[2] For these historians, as Will Mackintosh has rightly noted, "reliable and inexpensive transportation was a material precondition for the larger social, economic, and political effects that they trace."[3] Transportation itself was rarely the main event.

The historiographical terrain has shifted since 2010 as historians are no longer as transfixed with the market revolution, but the relative lack of emphasis on transportation outside of an economic context has continued. As historians have been teasing out the full implications of the relationship between capitalism and slavery, the literature has enriched our understanding of the economic links between the North and the South and the material conditions of slavery, sparking a vibrant public debate. Among these historians, however, transportation is still valued chiefly for its economic significance. In the work of Edward Baptist and Sven Beckert, for example, transportation is part of the necessary backdrop to their larger stories of economic transformation in the antebellum era. Transportation plays a larger role in the work of

Calvin Schermerhorn, who chronicles how steam transportation assisted the internal slave trade in his study of the "business of slavery." Steamboats play a prominent role in Walter Johnson's history of the Mississippi Valley. While charting the cotton kingdom, Johnson also pays attention to the wide-ranging cultural impact of steamboats, on the sensory experience of travelers, explosions, gamblers, races between boats, literature, and a host of other topics.[4] For many historians, however—regardless of the larger paradigm in which they are working—such nuanced treatment of transportation is an exception. Rather than a central focus of history, transportation often remains the backdrop against which other changes played out.

In this book, I argue that we should take transportation out of the backdrop and bring it front and center. In so doing, we can discover that it has much to teach us. The mechanisms by which economic change accelerated in the antebellum era—the application of steam power to transportation—effected a social and cultural transformation in the country. Steam transit constituted the first *direct* contact that many Americans had with the transformations wrought by the development of steam power.

To be sure, Americans had frequent contact with *products* of technological innovation: they devoured books and newspapers from the printing press or bought clothes made in steam-powered mills. But transportation improvements were different. When people boarded a steamboat or a railroad, they were no longer one step removed from technological power. Rather, they delivered themselves directly into its hands. No longer were they responsible on an individual level for their own safety, as they might be when walking or riding a horse. Nor would accidents affect only a small group of people, as they might in a canoe or a stagecoach. Instead, passengers entrusted their lives to the crew of strangers on a steamship or a railroad, and hundreds could be injured or even killed in an accident.

Americans were acutely aware that they were embarking on a dynamically different relationship to technology. Traveling by steam transit made Americans feel like they were part of the most modern, exciting technological development of the age, even if they did not fully understand how it worked. The finer points of the technology probably eluded most passengers. As David Nye has noted, a "passenger on a steamboat or a train comprehended it first as a novel experience of noise, power, and movement, and later as a dynamic part of a larger narrative about American expansion and progress, but probably never in terms of thermodynamics."[5] Even if most Americans never grasped how steam engines worked, they recognized that they were participating in a cutting-edge and exciting technological development with wide-ranging impact.

To learn how Americans understood these technological developments, we need to turn to the cultural realm. *Culture* can be a slippery term, and in the context of steam transit, culture could be many things. In this book, *culture* will be used chiefly in two ways. The first use of *culture* is to indicate a set of practices or expectations. As steam transit grew, suppositions about behavior on board railroads and steamboats developed alongside it. Some of these ideas were communicated by etiquette guides—published works that allowed novice travelers to learn how to behave. Other aspects of this culture were created on the fly by travelers themselves, as when rules for claiming a vacated seat were enforced not by the corporation or etiquette authors but by other passengers. Thus, this culture was both something proposed by elites and a series of rules created democratically.

The second use of *culture* relates to how steam transit influenced the creation of popular, artistic, and educational artifacts consumed by Americans. It is perhaps easy to see how railroads and steamboats influenced popular culture: their images appeared in books and on currency, songs were written about railroads and steamboats, and novels and short stories took transportation as a stage for exciting action. All of this culture could be consumed far from the tracks or the docks. But people also imprinted cultural relevance onto the infrastructure of steam transit, as when a railroad depot was commandeered for a church service, musical performance, or political meeting. Steam transit inspired cultural production and was in turn impacted by the cultural changes taking place. Cultural artifacts did not simply reflect back onto steam transit but helped reshape it in an ongoing process throughout the antebellum era. "Culture" could be both cause and effect. Through this extended cultural conversation, steam transit was "naturalized" for Americans like Sallie McNeill.

By arguing for greater attention to the impact of the Transportation Revolution beyond politics and economics, I contribute to a growing literature that strives to make the Transportation Revolution more three-dimensional.[6] Thanks to steam power, travel in the early nineteenth century was very different from travel on the eve of the Civil War. Prior to the nineteenth century, most travelers had to plan their own routes and might not even be certain where they could find food and lodging at the end of each day. But with the Transportation Revolution, travel became commodified. Traveling by steam, travelers simply purchased tickets and went on their way. Travel was something to be bought, like a hat, a newspaper, or a loaf of bread. Naturally, how a person experienced that commodification could change based on their race, age, class, geographic location, or gender, and this book explores those different experiences.

One final benefit of examining the social and cultural impact of antebellum transportation is to uncover the roots of stories told about the post–Civil War era. Historians have made great strides in broadening the history of steam transportation in the late nineteenth century. These studies have expanded our understanding on a diverse range of topics of how railroads related to gender, Native Americans, tourism and the landscape, Black laborers, consumerism, architecture, Ku Klux Klan violence in the Reconstruction South, and other topics. The rich postbellum social and cultural history described so wonderfully by these historians and others had antebellum antecedents that deserve fuller explication.[7]

While historians tend to treat steamboat and railroad travel separately, a quick perusal of antebellum travel accounts shows that antebellum Americans themselves moved frequently among different modes of travel. Therefore, this book will consider both steamboats and railroads together, under the rubric of *steam transit*. There is no question that as the antebellum era went on, railroads came to dominate the cultural conversation. Yet steamboats never disappeared, and they continued to make a major contribution to the American economy. Indeed, national steamboat ton-miles increased throughout the nineteenth century, even after the Civil War. Railroad companies and steamboat lines often cooperated in order to increase the number of routes, and thereby destinations, available to their passengers. Therefore, while there are times in this book that railroads may predominate, I have worked to include steamboats as well, since they were always a critical part of the antebellum transportation menu.[8]

When confronted with the wide array of places in which steam transit influenced American life, a historian has a number of different paths to consider. I have selected seven topics that I argue allow for a richer understanding of the social and cultural history of the Transportation Revolution. Throughout the text, these examples are drawn from all areas of the country, demonstrating that this cultural and social change was happening across multiple boundaries: North and South, enslaved and free, urban and rural, and so on. Naturally, the scale of impact could be different in different places, and the pace of change moved unevenly across the country. When McNeill wrote in her diary of her first sight of a railroad, five states already had more than two thousand miles of track within their borders.[9] Although the experience was *common* across the country, it was not necessarily *unifying*. Despite the frequent booster rhetoric of railroads binding the Union together, steam transit could not prevent disunion in 1861. Some passengers blanched at the thought of sharing a seat or a deck with someone from a lower class. And Black passengers, enslaved and free, suffered separate treatment in both the North and South. There was no singular travel experience on steam.

This book's seven chapters may be grouped into two parts. Chapters 1–4 consider how steam transit interacted with the broader culture, starting with its construction and then moving to the travel experience, the arts, and religion. In chapter 1, I explore the relationship between transportation companies and the communities they served. At the very beginning of the Transportation Revolution, steam power was new, and it had to be explained to communities to secure their support. Transportation companies had to do outreach to help residents understand what steam transit was and how it would operate in their neighborhoods. This was particularly true in the case of railroads. While steamboats brought a new type of motive power to an existing mode of water transit, railroads were something entirely new on the landscape. Railroad corporations had to enter into delicate negotiations with landowners in order to secure the right-of-way for the tracks. Landowners were generally favorable to railroads but did not want to give up something for nothing. Both sides in the negotiation looked to turn the desire for better transportation to their own advantage. As transportation grew, corporations fielded complaints from users and had to adjust to the demands of the markets they served. The chapter closes with an exploration of how transportation became more closely integrated into daily life in ways that the corporations themselves did not intend.

Chapter 2 looks closely at the experience of antebellum travel as described by the travelers themselves. Travel before steam transit could be arduous. Travelers had to assemble their own routes as they went and lodge where they could, and they were at the mercy of the elements if they were traveling by horseback or on foot. The first steam journey, then, was frequently worthy of comment in diaries and letters, and the chapter examines how people reacted to their first trip aboard steam transit. As steam transit became more regular, social norms developed about appropriate behavior on board a train or a ship. People wrote about these norms—and expressed their annoyance when the norms were breached. Steam travel invigorated all of the senses, and the sensory impact of travel attracted wide comment from travelers. The chapter concludes by considering the danger that steam transit posed to people's lives and attends to how Americans acclimated themselves to the risk of fatal injuries when traveling.

Chapter 3 looks at how steam transit found cultural expression in the arts. Steam transit's movement into popular culture demonstrates how economics alone is insufficient to understand steam's impact. Steam transit lent itself to an abundant number of metaphors or other verbal expressions, which quickly became widespread. Writers in national publications and ordinary Americans in their private correspondence used metaphors of steam in their writing. Humorists trained their sights on steam transit, initially offering comically exaggerated assessments of its performance as a way to skewer the relentless boosterism of the age. Later, as steamboats and railroads became more common throughout the country, humorists mocked those unfamiliar with steam transit as backward or out of step. Music composers commemorated construction of new railroad lines or steamboats or attempted to mimic the sounds of steam transit in their own compositions. Dozens of songs were composed in the antebellum era that people could play at home on their own pianos or hear when they visited a friend's parlor. And a wholly new genre of literature sprang up around steam transit as guidebooks gave travelers tips about the best way to travel. For those who remained at home, these same guidebooks painted a picture of what could be seen along a given route, allowing readers to visualize the trip from an armchair.

Chapter 4 surveys the reaction of Protestant ministers to the development of steam transit. Many saw in steam transit a powerful sign of God's favor toward the United States. They recognized that this technology could be a tool to carry the word of the Gospel to more people more quickly than ever before. Ministers and others also worried about the moral damage being done to workers and passengers alike when transportation operated on the Sabbath. In an example of how culture pushed back against technology, the Sabbatarian movement advocated strongly against Sunday operation, with some success. The fact that any trip could end in a deadly accident meant that Americans turned to their religious leaders for answers and meaning. Accidents on steam transit robbed Christians of a "Good Death," dying at home surrounded by loved ones and at peace with God. Ministers responded by preaching that good Christians had to live a faithful life and be prepared that life could end at a moment's notice.

The second part of the book consists of chapters 5–7, which consider how three different groups in antebellum America reacted to steam transit. Black passengers, as explained in chapter 5, experienced transit in very different ways in the North and the South. Enslaved labor was critical to building the southern railroad network, and after that network was constructed it both helped expand slavery by facilitating the internal slave trade and served as a tempting means of escape for enslaved people. In the North, railroads and steamboats alike formed a critical portion of the Underground Railroad, as Black abolitionists and their white allies recognized the value of moving people as quickly as possible to the Canadian border. Northern railroad depots and steamboat wharves became sites of public confrontation when Black people were threatened with re-enslavement by those who attempted to transport them back to the South. And for free Black people in the North, treatment aboard steam transit could be demeaning, as inconsistent treatment at the hands of white employees and passengers alike challenged their rights and their dignity. Thus, Black people were both voluntary and involuntary travelers on steam transit in the antebellum era, depending on the time and place of their travel. The ingenuity of enslaved people in making use of steam transit stood in contrast to the increasingly commodified travel available to white Americans.

For white women passengers, the subject of chapter 6, transit opened up new possibilities for mobility. Steam transit represented a complex space for women—neither wholly private nor public—which meant that there was much to consider for women travelers. Scores of etiquette guides sprang up in the antebellum era, and they counseled appropriate decorum for women, for when women did travel, they were thrust into a world of strangers. For some women, traveling with a male chaperone, who could contend with the tickets and the baggage, ensure that the proper transfers were made throughout the journey, and prevent anyone from taking advantage of them, provided some security in addressing this new landscape. But other women chafed at the restrictions of a chaperone and saw steam transit as an opportunity to strike out on their own. Women's magazines also included fiction with scenes set on steam transit, helping familiarize women with the changing world of technology.

Finally, chapter 7 assesses children. Like Sallie McNeill, these children were born into the world that steam transit was changing. They grew up with the developments in transit and did not know any world other than that which steam transit defined. For the adults raising these children, there were urgent moral questions to be addressed in a changing world. The antebellum period was one of intense social dislocation: white Americans began to move west, in the process breaking up families and throwing into disarray formerly stable relationships. For many who remained in the East, the transition from agricultural to industrial labor and the growth of wage labor called into question more traditional conceptions of self-sufficiency. Technological progress during the antebellum era heightened the moral questions faced by parents, but technology also offered solutions. Namely, the growing children's literature during this time period used steam transit to help teach moral lessons. A steamboat journey or railroad trip could serve as the centerpiece for a novel or a story and in so doing teach the young reader critical lessons about responsibility, independence, and morality.

We cannot know precisely what Sallie McNeill "read and heard," but this book will demonstrate that steam transit thoroughly permeated American cultural life, and in turn American culture pushed back on technology. Steam transit transformed the antebellum world and was transformed by that world as well. The wide range of cultural adaptations inspired by railroads and steamboats speaks to their overall force. Steam transit was not developed with religion, music, humor, or any of these other cultural aspects in mind. Railroad boosters may have nodded rhetorically toward a generalized notion of progress, but their eyes were chiefly on economic improvement. McNeill, by contrast, provides one such example of how the rich cultural experiences informed by steam transit could prepare people for in-person experiences with steam. As the machine was naturalized in American life, her experience was replicated countless times across time and space in the antebellum era.

CHAPTER ONE

Community Relations

Steam transit offered possibilities beyond what human, animal, or wind power could offer but also required Americans to accept alterations to the natural landscape and soundscape. Steamboats operated on existing waterways but were noisier and more dangerous than other watercraft. Railroads were even more intrusive: they could not operate on existing paths but had to create their own pathways, leveling the land before them. Given the changes that steam transit wrought, steamboat and railroad corporations had to work with the communities in which they hoped to operate. Sometimes this work was cooperative. Most communities welcomed steam transit, although that enthusiasm could be tempered once they became more aware of its costs. In other cases, corporations and communities were more antagonistic from the outset. Sometimes the corporation's desires won out; other times the communities pushed back. By the end of the antebellum era, the infrastructure of steam transit was widespread, and communities were exploiting that same infrastructure for their own ends, even in ways that the corporations did not intend or foresee. In so doing, different parts of the community pushed beyond strictly economic or political considerations and expanded the cultural use and purpose of these technological changes. Such interventions and adaptations demonstrate that steam transit did not have a one-way impact on culture—others adapted technology for their own use.

Corporations learned quickly that they needed to work with the communities in which they hoped to operate. This meant that corporations had to place a priority on education: railroads and steamboats were new, so the corporations had to teach people what steam transit was, how it worked, and what benefits it would bring. Corporations also approached their work with a spirit of experimentation. What may seem now to us like obvious realities of steam transit were open questions in the early nineteenth century. Would railroads operate as open highways, where anyone could bring carriages onto the tracks, or would only one company be allowed to run trains? Would trains be pulled by horses or steam engines? If two trains met going in opposite directions, how would precedence be determined? All of these questions, and more, were undecided at the beginning of steam travel, and answering them required experimentation. The nineteenth century landscape saw both successful and unsuccessful experiments, and the pioneers in this technology learned from both.

Corporations built their networks in communities, which meant negotiation and conflict with individual landowners. These individuals pushed back, hoping to keep their own rights and advantages. Such negotiations, replayed across the entire country and across decades as infrastructure expanded, represented the first nexus of community engagement for companies. While generally excited about the improvements that steam transit would bring, community members did not hesitate to protest against mistreatment by corporations or their representatives. To protect themselves and set expectations for performance, corporations instituted policies to determine when they would and would not be responsible for accidents or damage to goods. In hearing and combating complaints, railroad companies began to answer some of the open-ended questions that were posed at their creation, and in so doing, they defined the limits of their power. Finally, community members did not passively accept the changes in the land wrought by steam power but also worked to mold it to their own ends. In the history of technology, historians have been keen to examine technologies for ways in which individual users modify and repurpose technologies for the users' own purposes.[1] Steam transit was not a technology that users could bring into their own homes and modify. But plenty of people found alternate uses for technology that expanded the utility of steam transit far beyond getting from point A to point B.

In this way, the users of technology—the members of the community—expanded the cultural reach of steam transit by enlarging its potential uses beyond transit. Someone walking by a train depot might remember not just an important trip but potentially also the time they saw the famous soprano Jenny Lind perform at that depot. Or they might recall when the town was so new and its infrastructure so skeletal that a railroad car was pressed into service as a Sunday school classroom. Or they might recollect the time an angry crowd formed to turn back a southern posse attempting to capture a Black man and carry him to slavery. As transit moved from a new thing on the landscape to a natural part of the landscape, communities gave cultural meaning to this corporate infrastructure.

The Novelty of Steam Transit

"What is a Rail-Road?"

With this question in 1827, a railroad advocate in North Carolina underlined how novel the technology was for most of his readers. People had to be instructed about what railroads were. "As many persons have not had an opportunity of knowing the manner of a rail-road," the writer continued, "it will be well to give a description of it." But the author assured his audience that it need not be concerned about the complexity of the technology: "It is so simple in its construction, that any one will easily understand it." He then proceeded to spell out how the track was constructed, how the cars of the train would be prevented from leaving the tracks, and other crucial aspects of railroad operation.[2]

Clearly, the railroad had not yet achieved widespread familiarity: its basic construction and operation needed to be spelled out for ordinary Americans.[3] Whereas steamboats constituted a new form of power applied to familiar water transit, railroads—with their tracks, engines, stations, and rules for precedence when sharing a track—were venturing into something new. Railroads eventually became an accepted part of the landscape, but boosters saw public education as necessary to achieve that end. In order to get the public's support for railroads, they wanted the public to understand how they operated. This satiated the public's curiosity for information and demonstrated that the new method of transit was entirely safe. In 1835, a railroad convention in Dover, New Hampshire, opened its report with a definition of a railroad: "RAIL ROADS are contrivances for obtaining surface for the wheels of carriages smoother than the surface of a turnpike or common road." The report continued to lay out how a railroad was constructed, how cars would be prevented from sliding off the tracks, and how trains would pass each other on a single track.[4] Some railroad descriptions used the familiarity of steamboats to reassure people about the newer form of steam transit. For example, one advertisement for a particular type of railroad carriage declared that in this carriage travelers aboard could expect "all the convenience and comfort which belong to the best steam boats."[5] The steamboat may have been brand new just two decades before this broadside was released, but in the intervening time it had become a comfortable point of reference for an even newer technology.

Boosters did not have to work hard to interest the public in these works. When the technology was new, it elicited tremendous curiosity. A writer in 1823 illustrated the heightened expectations that came with internal improvements. "Rail roads, Locomotive Engines, &c &c are all the go here," Charles Endicott wrote, describing the advocacy around the Baltimore and Ohio Railroad. Turning slightly tongue-in-cheek, he continued, "The poor are to be made rich, the miserable, happy, the naked clothed, the hungry fed. In short there is to be a general revolution in men, manners, and things, in this part of the country. And all this is to be done by the Baltimore rail road." Although Endicott was exaggerating what the railroad would accomplish, there is no doubt that all of these things were positive, and turning serious again, Endicott averred that the railroad "cannot fail to benefit our city in a great degree."[6]

The public interest in improvements was genuine. People noted in their diaries when they saw steam transportation for the first time or that they made a special trip just to observe a steamboat or a railroad. On June 5, 1815, Harriott Pinckney Horry arrived in Georgetown (now part of Washington, DC) via steamboat and noted that "all sorts of people were flocking to see it," since it was "the first Steamboat that has appeared in these parts."[7] On May 13, 1822, Connecticut resident Nathan Foster wrote that he "rose early, & walk'd with Mrs. F. to see steamboat," underlining the particular object of interest.[8] In 1835, Caroline Poole, in Massachusetts, noted that she took a stagecoach to Lowell and then after tea "walked down to see the cars come in."[9] The arrival of the railroad marked a special event worthy of her attention and notation in her diary. People continued to note this type of interest into the rest of the antebellum era. In 1841, nineteen-year-old Evan Randolph took a stage journey through western Massachusetts. After leaving the stage, he and another passenger "walked a mile to see a cut 70 ft deep through rock" to accommodate the Western Railroad. Randolph marveled that they saw a steam shovel, which could "dig out a cart load of earth every few seconds."[10] Even before the railroad was operating, the construction of the railroad itself was a technological wonder that Randolph made a special effort to see. Animals also noticed the new machine. When Elizabeth Steele Wright's dog, Growler, saw a train for the first time in 1848, he "barked" and did "all he could to frighten the engine."[11] Foster, Poole, Wright, Randolph, and Growler alike demonstrated the widespread interest in new technology.

Steam transit also earned mentions in the popular press. Newspapers covered steamboats and railroads extensively; for example, historian Craig

Miner estimates that he examined around four hundred thousand articles about railroads from 185 newspapers for his 2010 book.[12] Beyond simple reporting, steam transit also attracted attention in the "newsboy's appeal"—a long poem generally published at the end of the year summing up the year's events, which newsboys hoped would prompt a gift from the newspapers' readers.[13] Given that railroads and steamboats were frequently in the news, they in turn featured prominently in these appeals. In 1829, one appeal noted that railroads were *"long talked of,"* signaling that they were part of general conversation in the town.[14] Another early appeal made the comparison of steam travel to an older method of transport, the horse: "You heard how many miles an hour, / They go by steam and rail-road power. / If we such a machine had got, / You'd have your papers smoking hot; / 'Twould save some birch on our old Bob, / And poor old Dobbin many a job!" Bob and Dobbin are presumably names for horses, with "birch" a reference to the switch used to spur the horses. The newsboys envisioned steam transit as a replacement for horses, which would both be speedier and spare the horses a whipping.[15]

Another sign of popular enthusiasm could be seen from an appreciation of steam transit's aesthetic appeal. A real estate advertisement from 1836 in Worcester, Massachusetts, included railroads as a selling point for the land not for economic reasons but because the railroad was pleasing to the eye. The property was described as having "a view of the cars on the Boston and Worcester Rail Road, and of the boats upon the Blackstone Canal, for a considerable distance, and will have a full view of the cars upon the Norwich and Western Rail Roads, for several miles."[16] While the potential purchaser of the lots might imagine traveling to a new location or the goods that could be brought in by the railroad, the chief selling point here appears to have been the very existence of the railroad: something to observe and admire from a distance, not merely to get from one place to another.

Perhaps the most significant sign of popular enthusiasm for railroads was the excitement that accompanied a groundbreaking, or opening of a newly completed section of track. In such events, the corporation and the community came together to celebrate—the most publicly visible sign of unity between the two. Descriptions of these events abound in the antebellum era. One of the most complex was the gala celebration at the groundbreaking of the Baltimore and Ohio Railroad in 1828. Dozens of workingmen's groups participated, choosing to imbue the groundbreaking with their own particular message. The celebration attracted farmers, bakers, ornamental chair painters, tanners and curriers, bookbinders, watchmakers, glass cutters, rope mearers, draymen, and many, many others. In a long parade, each workingman's organization processed with a demonstration of their artisanship, or was costumed in such a way that praised the prosperity of their country, applauded their own individual type of work, and lauded the railroad. The masons and bricklayers, for example, carried three banners. One banner showed "a house partly built, men at work, &c.," thus demonstrating the value of their trade—providing shelter. The inscription at the top of the banner read, *"Masons and Bricklayers of Baltimore, united July 4, 1828"* and at the bottom *"Liberty throughout the world,"* tying their work to larger themes of independence. Finally, the masons were dressed in "aprons ornamented with the emblems of their profession; their badges had on them a trowel, and a representation of a rail road."[17] The masons were declaring the importance of their own place in society. They did not merely celebrate the railroad but incorporated it into their own story.

During the parade for the Baltimore and Ohio Railroad, several of the groups performed aspects of their profession displaying for the audience their own abilities and skills. The cordwainers, for example had a stage drawn by four horses, which held "two master workmen, two journeymen, and two apprentices, engaged at work upon a pair of green morocco slippers, which were finished during the procession, and presented to Mr. Carroll on the ground. The slippers were very neatly made, and the linings were ornamented with a view of the Rail-Road."[18] The cordwainers thus had representatives from every stage of their profession on display, working together to make the slippers. Mr. Carroll was Charles Carroll of Carrollton, the last living signer of the Declaration of Independence, who turned over the first shovelful of dirt for the Baltimore and Ohio Railroad's construction. This parade demonstrated that the railroad was not just about getting from one place to another. Rather, it was something for the entire community to celebrate and a platform for laborers to extol their own achievements and independence as working men. Even groups like ship captains, who had good reason to be concerned about possible competition from the railroad, participated in the parade. In their display, the ship captains praised the railroad "as an instrument of urban growth, one that would enhance, not supersede, the city's maritime interests."[19] In a parade and celebration such as this, different groups of Americans imprinted their own meaning and message onto the railroad's construction, separate from what the corporation itself was interested in doing.

In this same vein, local communities continued to celebrate the completion of the railroad, even after the technology was long familiar. Numerous examples highlight the sustained excitement that railroads brought to a community throughout the antebellum era. For instance, hundreds of people attended a barbeque when the Richmond and Danville Railroad was extended to Amelia Court House, Virginia, in 1851. In 1853, citizens boarded a special train to celebrate the completion of a section of the Virginia and Tennessee Railroad, and the festivities concluded with a meal.[20] A railroad groundbreaking in Florida in 1856 attracted "representatives of the Masonic Lodges, the Odd Fellows, the Fire Company, and the Temperance Societies as honored guests," and a band "led a parade" to the "site of the future depot."[21] Finally, in Memphis in 1857, thirty thousand people gathered to celebrate the completion of the Memphis and Charleston Railroad. Muddy streets made walking difficult but did not prevent a parade two miles in length stretching through the town, followed by speeches, a dinner, and "pouring into the Mississippi River a barrel of salt water brought from the Atlantic Ocean by the Charleston delegation."[22] Memphis's population in 1858 was only twenty-two thousand, so even if this report of the celebration overstates the exact number of people present, it surely reflects the wide popularity of the event.[23]

Public exuberance, parades, and picnics did not guarantee immediate success of a railroad, which still had to overcome skepticism. Much of the best-known early skepticism was articulated by artists and writers. For instance, artist Thomas Cole lamented that the construction of a railroad had ended the "charm of solitude and quietness" that had once been the characteristics of his "favorite walk."[24] Some notable American writers obtained a reputation for opposing steam transit. Henry David Thoreau, for example, famously wrote that "we do not ride on the railroad, it rides upon us." But even these statements did not reflect a wholesale rejection of technology. As historian Matthew Klingle reminds us, Thoreau also wrote, "I am refreshed and expanded when the freight train rattles past me," since it carried with it produce from all over the country and made him "more like a citizen of the world." Klingle argues that Thoreau "hated the railroad because it reminded him of his failure to live independently."[25] For writers such as Thoreau and Ralph Waldo Emerson, the fear was less of the railroad itself than of "the subordination of imagination to materiality"—that people would abandon their creativity in the face of the pursuit of the material gain made possible by the railroad's easy provision of ready access to markets.[26] As the examples of cultural production considered later in this book make clear, Americans did not give up their creativity in the wake of steam transit.

In local communities, the objections tended to be less conceptual and more practical. One early question was whether railroads would operate as "common highways," open to anyone who wanted to use them, or be limited to rolling stock controlled by a single company. Some communities wanted corporations to provide open access to the tracks, rather than letting all of the benefits fall to a single corporation. But when this experiment was put into practice, it quickly proved to be unworkable. In a memoir, engineer Solomon Roberts recalled the challenges he faced when a railroad operated as an open highway: "The drivers were a rough set of fellows, and sometimes very stubborn and unmanageable." Specifically, he recalled one driver who "would not go backward, and could not go forward, and so obstructed the road for a considerable time." The company could do little in such cases of hard-headed drivers. Another problem with running the railroad as a common highway was determining who had right-of-way on a single track. Roberts remembered that his company put up a post, equidistant between two turnouts, and whoever reached the post first would have precedence. Drivers quickly realized this and adjusted their own speed to game the system: "When a man left a turnout, he would drive very slowly, fearing he might have to turn back; and, as he approached the centre post, he would drive faster and faster, to try to get beyond it, and thus drive back any cars that he might meet."[27] As drivers attempted to beat the odds, accidents, arguments, and standoffs were the inevitable results.

For that reason, railroad corporations moved relatively quickly to establish that one company would control all aspects of travel on a given road. In 1838, a pamphlet gathered together the objections of civil engineers and others as to why multiple corporations should not operate on the same road. One chief engineer wrote, "The great speed and irresistible momentum of these machines and their trains, render any other than a unity of management in the highest degree dangerous, if not absolutely impracticable." Others agreed; another engineer wrote, "The advantages of rapid travelling on Rail-Roads now obtained, would be very much reduced, if not wholly destroyed by the admission of other engines than those belonging to the owners of the road on which they are used."[28] For reasons of speed and safety, engineers urged single control of a railroad track, and that soon became the common position. Running the railroad as an open highway was an experiment that proved too risky.

Railroad companies acknowledged that such experimentation, failure, and reevaluation were necessary for progress. In its annual report released in 1843, the Western Railroad acknowledged that the previous year had been "emphatically one of experiment." Nevertheless, the company stressed, "It cannot be doubted, and it would be surprising were it not so, that in a business so untried, expenses have arisen and difficulties have been encountered, which might have been avoided, by the experience and knowledge now possessed, but by these only."[29] With this new technology, only by trial and error could the company gain the knowledge it needed. Likewise, the Little Miami Railroad (of Ohio), in 1848, informed its shareholders that the work had continued even though the company had been from time to time "embarrassed by the novelty of the enterprise, and the inexperience of all connected with it."[30] Experimentation and failure were an inescapable part of early steam development, and corporations urged their shareholders to show patience and accept that setbacks were inevitable.

Negotiation between Companies and Communities

The need for and degree of community interaction differed for railroads and steamboats in part because each mode had a different way of interacting with the natural world. Steamboats could operate on existing waterways—created by nature—and could easily share rivers with other private watercraft. Open waterways allowed for greater maneuverability, and boats could work their way around each other if need be. Railroads, by contrast, could not rely on existing avenues, and the design of railroads meant that only one train could travel on a segment of track at a time (unless a corporation was willing to spend money on an expensive second track, which few companies were willing to do). Railroad corporations had to acquire land and build their infrastructure, which meant that they were inevitably negotiating with existing property owners.

The prospect of building a railroad excited talk and speculation in local communities. In part this was because the exact benefits of the railroad were unproven. William Hasell Wilson, a railroad engineer, worked on the railroad from Philadelphia to Columbia, Pennsylvania, in March 1828. He reported that most farmers along the route "had no faith in its accomplishment" and worried that if the railroad were completed, "the adjacent country would be ruined, as the city market would be overstocked with agricultural products from a distance, where land was cheaper and expenses less." Wilson noted that farmers also feared that the railroad would lead to a decrease in traffic from wagon-drivers, who purchased feed for their horses from farmers as they moved from the inland to Philadelphia. Wilson recalled that farmers were skeptical of the railroad even though at the time the plan was to use horses, not steam, to drive them. Even if horses rather than steam engines were used to pull railroad cars, farmers worried, fewer horses would be required than before, leading to correspondingly lower sales of feed. Wilson credited this particular fear with leading to the palpable opposition of many farmers to the railroad.[31]

When seeking the most advantageous routes for their tracks, railroad corporations tended to define as "natural" the inherently artificial structures they wished to build. There is no doubt that in pressing the "natural" advantage of their preferred route, railroad advocates hoped to convince communities that the route was preordained. As historian David Schley has pointed out, "invocations of nature masked will."[32] By arguing that a certain route had a "natural" advantage in trade, town boosters and companies hoped to make seem inevitable what in reality could be a fraught process of determining where tracks should run.

Anecdotal evidence from the time suggests that most landowners across the country did see an increase in land value if a railroad was constructed adjacent to their property. In 1852, Chastina Rix recounted that her uncle in Vermont was able to sell his farm for $2,000, an increase of $1,400 from its previous value, thanks to the nearby location of the railroad.[33] That same year, Lincoln Clark wrote that he was pleased with his purchase of land near Dubuque, Iowa. Thanks to two railroads, both in "rapid process of construction," Clark was confident that Dubuque was "bound to be a place of importance." He concluded by telling his wife, "I do not think we could have made a better location—I am perfectly satisfied with it."[34] This anecdotal evidence has been supported by more rigorous analysis from modern scholars. Looking at land values throughout the United States, Lee A. Craig, Raymond B. Palmquist, and Thomas Weiss found that access to rail increased land values "by about 4%." To be sure, these authors found that farmers could also benefit from access to the oceans or Great Lakes. But whereas the Great Lakes and oceans are fixed in place, "improvements in river transportation, or the construction of railways or canals . . . were possible in many locations."[35] Thus, while any access to transit was good for the economy, steam transit was more widespread, could expand, and could benefit more people.

Historical studies focused on smaller communities have found similar responses across the country. One study of Knox County, Illinois, found that with increased distance from the railroad, "the value of land dropped off quickly and leveled out rapidly."[36] Thus, land nearer the railroad was more valuable. Historian William Link notes that in southwestern Virginia, "the 'very anticipation' of the completion of the Virginia and Tennessee Railroad increased local property values by a third."[37] In Edgefield and Barnwell Districts in South Carolina, "pine barrens that had sold for less than fifty cents per acre vaulted to between $1 and $5 per acre after completion of the Charleston-to-Hamburg [railroad] line."[38] And in Georgia, the construction of the Western and Atlantic Railroad led to "six out of the ten counties along its path" having the "highest average value of land per acre in the state" in 1860.[39] Across the country, transportation improvements demonstrated their economic benefits to local landowners.

With such economic advantages, some landowners were anxious to have the railroad run near them, in order that they might profit from the road. Such landowners attempted to influence the route. When working on a railroad in Minnesota, civil engineer Walter Gwynn Turpin wrote to his wife that he was "constantly" approached by people who were "trying to buy me up to run the road this way and that way, make the depot here and there, all making large promises. Many offer to advance money in any sum for me to purchase lands on line of the road offering me half the profits." He claimed that he was not bothered by all the attention, and he treated the petitioners equally: "I give all civil answers, nothing more." Nevertheless, the persistence with which the Minnesotans pursued him was striking: "The rascality out here surpasses any I have ever encountered anywhere."[40] Clearly these Minnesotans were keen to have the railroad built near them and to profit from it.

Publications touted the rewards that steam transit would bring, especially in the realm of commerce. Steam transit's advantages were democratized, whereas before, transportation at scale was the province of elites. For example, for a plantation owner in the eighteenth-century South such as Henry Laurens, the critical need to move commodities to market meant developing his own "fleet of schooners" that he could send out on the local waterways when necessary. Laurens prided himself that he could send these boats when it worked to his own advantage, "without dependence on any intermediate Market."[41] Steam transit brought similar access to all, without having to purchase one's own individual fleet. In 1844, the *Maine Farmer* mused that the market for mutton would improve, since "steamboats and railroads have, in effect, moved us very nigh Boston."[42] The following year, the editor of the *Farmer's Monthly Visitor* took a journey through New York, Massachusetts, and Connecticut. He noted that the ability of farmers to supply Boston with milk was vastly improved, since farmers could more quickly get the milk to

Boston before it spoiled. In the 1810s, any attempt to transport milk over existing roads would have "churn[ed] the whole, before its arrival, into sourness and buttermilk." By contrast, on the railroads distant farmers could supply milk "with equal freshness and facility as that of the milk farmer less than half a dozen miles out of Boston."[43] By decreasing distance and enhancing the quality of the journey, railroads democratized possibilities for both farmers and consumers, and neither had to be independently wealthy to reap the rewards of improved transportation.

Despite these advantages, railroads could present challenges to the livelihoods of those who lived along the land as well. Tillable land was lost to the track. Tracks could cut off one portion of a person's property from another portion; once-uninterrupted land was now broken up by a dangerous crossing. Elevated tracks with ditches could change water flow and cause erosion. Sparks from an engine could trigger fire. Construction crews could damage land, even that which was not given over to the railroad. Trains could frighten livestock, and unfenced track could threaten the lives of livestock if they wandered onto the track. Landowners along the routes of railroads hoped to gain just as Rix's uncle had in Vermont but also wanted to ensure that they received fair value for any potential damages such as those outlined here. The result was a complex land acquisition process.

Corporations realized that land acquisition, if dragged out, could seriously delay the already difficult process of construction. Land acquisition meant putting together an enormous jigsaw puzzle along the route of the railroad, with the hopes that no one individual piece would prove to be too costly. Therefore, companies wanted to move quickly; one railway agent in 1844 reported on his activity by noting, "Land holders called on as fast as possible."[44] Railroads also made it clear to their agents and engineers that they could not afford to be too lenient with the people who held the right-of-way. Charles Russell of the Franklin and Bristol Railroad in New Hampshire instructed one of his engineers to be "pleasant but firm. . . . If we yield to Ladd we must to Parker & others."[45] If landowners got wind that one of them (Ladd) had received favorable treatment, then Parker and the others would soon demand the same. In an 1849 letter, an employee of the Virginia and Tennessee Railroad attempted to advocate for landowners turning over their land without compensation by appealing to the long-term relations between landowner and corporation: "If the Landowners give the right of way it creates a permanent & reciprocal friendly feeling between the company and the people along the line of road." The increase in land value would more than compensate any landowners for the railroad running through their property. The corporation would do better, the letter argued, if it were running through "a body of its friends." Any demands for claims, which the author felt would necessarily end up in the court system, could only produce "lasting feelings of hostility on both sides." Far better for the landowners to freely give up their land and accept that their benefits would not come in the form of direct monetary compensation but in the overall improvement of the economy and "friendly" relations with the corporation itself.[46]

The Virginia and Tennessee Railroad's characterization of the relationship between corporation and landowner was, of course, completely self-interested, and it proved to be too sanguine. Controversies between landowners and corporations happened across the United States and even led to fights. Historian Michael Connolly found that in the late 1830s, surveyors of the Boston and Maine Railroad occasionally "clashed violently with local landowners and farmers."[47] More likely than violence, however, were lengthy legal battles. For example, landowner David McElwain battled the Western Railroad of Massachusetts for years for compensation for the loss of his tannery. Although we cannot know for certain how truly devastating the loss was since we only have his side of the argument, there is no doubt as to how persistent McElwain was in making his claims. In September 1840, McElwain wrote to the company to ask for consideration beyond what he had already been allotted. Evidently the construction of the Western Railroad damaged McElwain's tannery to the point that he felt he could no longer use it. Moreover, the creation of the railroad led to the decline of the neighboring turnpike, which deprived McElwain of the business of putting up boarders. After putting the case before the county commissioners, McElwain had asked for $1,400 in damages; the commissioners rated his loss at $300. He had actually received $319.50 and then sought an additional $630 from the railroad company to compensate for the loss of business.[48]

After sending his own petition in September, McElwain gathered nineteen signatures in another petition in October in support of his views. The language of the petition again appealed to the better nature of the railroad men, praising the company for showing "commendable liberality" in previous claims and hoping that the company would make a "just remuneration" for what was taken from McElwain.[49] Over a year later, McElwain was still attempting to get money from the corporation. He prefaced his new petition by noting, "I am in favor of the R. Road, and ever have been." To substantiate this claim, he noted that he attempted to make his claim for damages low, but when he discovered that the railroad would destroy his tannery, he increased the claim. He once again asked consideration for the fact that he was unable to support his family and that the lack of business was causing him to cut substantially into his savings.[50] The directors of the Western Railroad remained unconvinced by McElwain's pleas. After attempting to reason with McElwain and to offer him additional money for his land, the company finally concluded that they were "under no obligation" to make any additional offer.[51]

J. C. Gray's negotiations with a railroad provide another example of how landowners sought to protect their investments in the face of railroad corporations' desired routes. In 1846, Gray entered into negotiations with the Fitchburg Railroad (of Massachusetts), protesting the fact that the railroad would bisect his land. Gray made his requests forthrightly: "In order to give me reasonable access to the land cut off," he wrote in 1846, "I think a bridge above the road for teams, will be required, near the middle of my North West field, and a culvert near the Northwest corner of my pasture, where the track first runs down to the pond." Gray did not want the tracks to cross his land "on a level." Moreover, the culvert that he wanted "should be large and high enough for loaded teams, whether for ice or otherwise." Finally, he demanded that "the fences, bridges, culverts &c to be of course kept in repair by the Corporation."[52] Gray sought a promise of maintenance from the railroad company, realizing that the building of the track was not the end of the relationship but rather the beginning.[53]

One month later, Gray continued to write about his frustration. The company clearly wanted him to agree to different terms, but he was not ready to do so. He wanted the company to select a route that was least damaging to his property: "The route now staked out would be far more injurious, in my mind, in every way, than the lower course first marked out on the shores of the Pond in the winter, & I should hope, that the Company might go back to their first purpose." Over a year later, Gray was still pressing his claims with the company: "I presume that the Fitchburg Rail Road Corporation are desirous, as I certainly am, to bring all questions between us to a prompt & amicable settlement. I have been waiting for the finishing of the fence & other fixtures on and about my land, which I regret to say are not yet completed." He then requested $4,000 for damages. The case appears to be finally resolved in January 1848, when a notation indicates that Gray received just over $1,800 in damages—a far cry from what he was asking for but perhaps more than the company wanted to spend, given how long they held out from making a resolution.[54]

Across the country, court proceedings could be expensive. The Pennsylvania Railroad discovered in 1850 that it did not allot enough money for land damages, due to the fact that "juries in most cases . . . exceeded our views of a just liberality in their awards in favor of the property holders."[55] Sometimes one court case led to others. A committee of investigation of the Boston and Maine Railroad outlined a series of pending court cases around land disputes in an 1849 report. In one case, arbitrators, "after an elaborate hearing," had determined the value of the land, which the company readily paid. But the landowner "refused to accept the sum awarded, or to execute and deliver the deed, or in any aspect to abide by the award, and has since formally revoked the authority of the arbitrators." The landowner then launched several suits against the company.[56] Such problems, multiplied along the length of the entire route, made acquiring land vexing for any proposed railroad.

Recalcitrant landowners could pose a problem even after construction began. Contractor Asa Sheldon recalled that in 1839, as he drove up with a load of construction materials to a site, a group of neighbors to the landowner drew their animal teams across the track to block his progress. Sheldon then "espied a woman hurrying across the field toward us, who proved to be the rightful owner of the land." She stood near the animals blocking the way, and since it was March the land was muddy and wet. Sheldon brought a plank for her to stand on so that she did not herself get wet. When another railroad employee asked Sheldon why he did not simply muscle the woman out of the way, Sheldon responded, "For more than twenty years I have not been in the habit of driving more than half way over so handsome a woman as that." This compliment "brought a smile to her face and loosened her tongue." Having prevented Sheldon's progress for thirty minutes, she relented and allowed the construction to go forward.[57] Railroad employees had to use all the tricks—and charm—at their disposal in order to get construction underway.

Individuals worried about the railroad on their private land, and communities also challenged railroad corporations regarding the placement of depots on common land. A railroad's location took on additional meaning because the presence or absence of a depot could spell economic doom for a community. As we have already seen, steam transit had the ability to alter the economic fortunes of one area in favor of its neighbors. Corporations realized the stakes of these decisions. Railroads represented competition between communities for economic advantage. "Commerce is a civil strife, and makes conquest of communities without the sword," noted the Philadelphia, Easton, and Water-Gap Railroad in an 1853 report.[58] Railroads could easily alter the fortunes of communities: the railroad could turn a once-prosperous area into something that was easily bypassed. Steamboats, too, could induce competition among towns. A decent stateroom in a steamboat could easily be preferred to a dingy tavern. Under these circumstances, the growth of steamboats spurred hotel-building as a way to make a town a better place to spend the night.[59]

As part of this fight to keep their own community's importance, communities were keenly interested in the construction and quality of railroad depots. The citizens of Springfield, Massachusetts, petitioned the Western Railroad about the location of the depot within that city. In so doing, they underlined Springfield's contributions to the line's success. They claimed that although the project may have owed much to the "wealth, intelligence & enlightened patriotism of the City of Boston," it was also true that the road was indebted to the "exertions & sacrifice" of Springfield residents.[60] Communities demanded the additional construction of depots if they felt the railroad was bringing them value. Inhabitants of Wilbraham and Ludlow, Massachusetts, also petitioned the Western Railroad, asking for the company to stop at Wilbraham. The petitioners argued that putting a station there would save residents of several towns "much trouble and time." Moreover, the students who lived in the area "would be better accommodated by being brought nearer the School."[61] Towns naturally hoped to entice the railroad through the promise of future business but, as we see, used other arguments as well, including time, convenience, and educational opportunity.

Railroad corporations ultimately wanted to make their decisions based on what was best for the corporation. Although railroads would listen to communities when it came to determining the placement of stations—particularly if such petitions were accompanied by promises of free land—one railroad president spoke for many when he claimed that the main object of the company was to place stations and turnouts as determined by "the convenience of the Publick, and the interest of the RR Corporation." For any other reasons than these, he argued, the company would not "vary the line one foot, to oblige a Town, Section, Corporation or even a Director."[62] By defining the corporation's interest as the public interest, the companies could pursue their own interest with the dressing of public benefit. In 1853, George Stark, the superintendent of the Nashua and Lowell Railroad, responded to people petitioning for a station by saying that while he would forward it to the board, "it cannot be granted without serious inconvenience to the majority of the travelling public," since the company already had "six stopping places on less than fifteen miles of road," which meant that the railroad could not operate "as fast as our competitors."[63] Stark highlighted the fact that the railroad was being judged on its ability to move goods and people quickly, and the demands of individual communities could not block that progress. These tensions between railroad company and community highlight the competing demands, varied uses, and perceived benefits of railroad development.

But corporations did not always hold the upper hand or receive a free pass from communities. In 1843, the Baltimore and Susquehanna Railroad reported that its ticket office in York, Pennsylvania, was in a tavern, and goods shipped by the company were held in a "rough frame building, the continued existence of which was objected to by the authorities and the citizens of the borough." Responding to complaints, the company built its own accommodations in York, and "with regret" the board had to authorize the expense.[64] In 1841, a subcommittee of the Executive Committee of the Albany and West Stockbridge Railroad (in New York) reported that finding land for depots was difficult. In each case, the land had to be acquired through "negociation," since there was "no power to take by process of law" the land that they needed. Moreover, in one of the locations, "the titles are so complicated & in the hands of so many parties, widely scattered that it has been found impracticable to get terms for purchase."[65] The quest to complete the jigsaw puzzle of land acquisition could be a difficult one.

Holding Corporations to Account

Once steam transit companies began their operations, customers held them to account and did not hesitate to complain about poor service. In order to ensure regularity and set expectations for performance, companies established rules and regulations for the handling of goods and passengers. This was true for both railroad and steamboat companies despite the different risks involved for each. In 1836, for example, the Portsmouth and Roanoke Railroad (in Virginia) announced its regulations for transporting merchandise. Its warehouse was open every day "sunrise to sunset, for the reception and delivery of Goods." People sending goods had the obligation to indicate "distinctly" on each item for delivery "the name of the consignee and depot at which they are to be left."

Certain articles were only "carried at the risk of the owner," such as "Glass, Stone or Crockery ware, Demijohns, and Jugs, either full or empty, and Furniture or Carriages not boxed, or Harnesses not securely bundled." Gunpowder had to be explicitly labeled as such or it would be refused passage. And sacks "containing Salt or any other article, must be strong and well secured, or they will only be taken at the risk of the owner."[66] In this way, the railroad attempted to accommodate a wide range of goods but also limit its own liability for mishandling en route. Other companies—such as the Petersburg, Greensville and Roanoke Railroad (in Virginia) and Raleigh and Gaston Railroad (in North Carolina)—established similar regulations for goods carried on those two roads in 1840. These two companies jointly declared that any item "of great value, such as gold, silver, &c. must be especially receipted for by the agents," or the railroad would not have any responsibility for what happened to it. And all items, once they arrived at their destination depot, would be left "at the risk of the owners."[67] The limitations articulated by railroad companies extended the negotiations between company and user, with both parties having to determine if the benefit outweighed the risks.

For their own part, steam transit customers wanted complaints to be handled expeditiously. Charles Eastman complained about some goods he received that had traveled by steamboat in 1828. He protested that the package he received was so wet "so as to damage nearly every article." He wrote with the experience of one who had suffered damage before, and he hoped to quickly resolve the claim: "I know not whether they were wet in the Steam Boat or since you received them, but should like to have you ascertain the fact as it can better be recollected at the time then some six months hence."[68] Eastman did not want to delay the addressing of his claim, but he wanted it handled while it was still in the memory of all concerned.

George Stark of the Nashua and Lowell Railroad (connecting New Hampshire and Massachusetts) referred specifically to the company's rules in 1853 when responding to a complaint. Apparently a customer complained about the treatment of crockery by the railroad. Stark acknowledged that "the best of servants are not at all times free from mistakes" but pointed out that the "loss of glass and crockery" could not be blamed on the company, since the company's "rules and advertisements expressly give notice that we carry and become responsible for, upon passenger trains, personal baggage only. As passenger trunks are of necessity handled somewhat roughly, our freight trains afford a much safer transport for such articles as you name, and it is only upon these trains that we undertake, and guarantee their transportation."[69] By placing the breakables in a passenger train rather than freight, the customer had taken all risk upon himself, in Stark's view. Companies set out rules for handling goods and then used those rules to defend themselves against claims from irate customers.

In addition to complaints from individuals, railroads continued to field complaints from the local communities they served. When railroads were built, they had every incentive to maximize use of their tracks, which could bring them into conflict with city residents. At the heart of these disputes was whether or not the railroad constituted a public function and thereby could interrupt the public streets. Groups in Baltimore protested further construction in the 1850s and thereby "rejected . . . the company's contention that facilitating rail travel served a public function, or at least the notion that long-haul travelers' interests in mobility outweighed their own right to move freely within their neighborhood."[70] In New York City in 1841, a committee investigated the possibility of removing rails already laid down by the Harlem Rail Road company, south of Fourteenth Street. The committee acknowledged that the railroad had brought benefits but that there are very different concerns in a "crowded city" than "when connecting distant marts." The committee concluded that there were "many evils which spring from the extension of the Harlem Rail Road south of Fourteenth Street." Therefore, the committee recommended that the rails be taken up.[71]

In Michigan, farmers along the route of the Michigan Central Railroad took direct action against the company when it began to pay out only one half the value of livestock killed by the railroad. The railroad management believed that it only owed one-half because the deaths were due to the farmers' "negligence and cupidity," but the farmers saw the payment of one half as "an admission of guilt, and bolstered by the state's open-range tradition, they continued to press the company for full restitution." The company would not budge, and so the farmers took matters into their own hands. In 1849, when scores of animals had been killed by the trains, farmers greased the tracks "with the lard salvaged from the carcasses of their dead livestock," presenting a significant danger to trains traveling on that stretch of track and symbolically using the train's victims to exact revenge on the train itself. Elsewhere, farmers at various times tore up track, burned down a depot, and even successfully derailed an engine. Although the perpetrators were eventually caught, the Michigan farmers' actions illustrate the depth of local concern about the railroad's impact on the economy and their livelihood. The tracks had an unforeseen impact on the more traditional ways in which farmers had managed and cared for their livestock. When faced with a challenge from the railroad corporation and inadequate recompense for the problems it caused, these farmers fought back in kind, with the goal of disrupting the economic engine built by the railroad corporations.[72]

In 1859, a group of concerned citizens in Richmond, Virginia, wrote to the Richmond, Fredericksburg and Potomac Railroad, hoping to block the purchase of a lot in town for the creation of a new depot. The citizens were property holders in the area and pointed out that building a depot in an area full of private residences "would not only injure our property in an incalculable degree, but would prove a public nuisance, by the din and clang inseparable from a Rail Road Depot, by the annoyance from smoke and dirt, by the greater liability of our property to be injured by fire, by the class of operatives which would be introduced among us, and by the danger to which our children would be subjected." The objections operated from a variety of levels: from the sensory disorder that would result from the depot (aural "din and clang" as well as the physical "smoke and dirt"), the danger of fire, the presence of lower-class workers at the depot, and the presumed danger to children from the activity in the depot. The citizens protested, "We purchased and improved our lots when no such nuisance existed in our neighborhood, and we cannot see that you have the right to injure our property without affording us any compensation, by erecting such an establishment in our midst." With that in mind, they promised to "use every effort in our power to protect ourselves from what we believe would be an infringement of our just rights." While the outcome of this particular protest is unknown, it demonstrates that the community did not hesitate to protest when it felt its rights were not being observed.[73]

Some towns placed restrictions on what railroads could do within their borders. In 1843, the town of Newburyport, Massachusetts, adopted by-laws that prohibited railroads from traveling faster than six miles per hour while in town, with a fine of ten dollars. Additionally, no railroads could "obstruct any street or side-walks of any street, for a longer space than three minutes at one time," with a fine of two dollars.[74] Likewise, Chester, Pennsylvania, prohibited the Philadelphia, Wilmington and Baltimore Railroad from running faster than five miles per hour within the town.[75] Charlotte, North Carolina, limited trains to four miles an hour in 1859, on pain of a fifty-dollar fine. The town added an aural requirement as well: "The bell shall be rung while running within the limits of said town."[76]

Sometimes corporations worked to resolve complaints amicably. The Boston and Worcester Railroad received a complaint that sparks from the engines were threatening nearby buildings with fire, and the board of directors responded by instructing the superintendent to "cause a netting to be placed over the funnel of each of the locomotive engines."[77] But corporations also fought back for their rights. A Worcester man, John F. Pond, complained to the city that the Boston and Worcester Railroad's trains were stopping unnecessarily on Summer and Front Streets. The railroad refuted the claim, stating that if the trains stopped, it was only for a moment to reverse the engine or throw a switch. The railroad protested that the city should not proceed with a legal case against the company, as evidently the city was ready to do. The company pointed out, "This road has always acted in good faith to the city of Worcester. . . . But it will not submit quietly to this kind of John F Pondism backed up by the city."[78] In this case, the corporation was willing to stand its ground and fight back against this citizen protest—cheekily calling out the opposing ideology by invoking the name of the protesting citizen.

Citizens angry at railroads could also take their case to the court of public opinion. In 1850, someone in Whately, Massachusetts, launched a newspaper to specifically protest the actions of the Connecticut River Railroad. The newspaper complained about the poor quality of the station house and against a superintendent named Hunt. One story noted that everything about town was "happy, prosperous, and joyful, save the old Station-house, which looks dreary still."[79] In a later version, the newspaper published a cartoon in which a figure sinking into the ground next to a dilapidated railroad station called out, "My friends help me out of this sand." A group of people, having built a side track to completely avoid the derelict station, responded, "HUNT your way out," a play on the name of the superintendent. The newspaper claimed that they wanted no more than their town was due: "What we want is justice. Do as well by us as is done by our neighbors and we will not have a word of complaint to make."[80]

As the preceding examples make clear, communities large and small across the country were anxious to see transportation expanded into the areas where they lived. Later economic analysis by historians has borne out what most local communities believed at the time: increased access to transportation had a positive impact on land values and the broader economy. But communities also demanded to be treated fairly, as they saw it, and were not interested in ceding total control of their land and livelihoods to corporations. When local communities did not believe they were being heard, groups did not hesitate to respond to the corporation with direct action. In all, the interactions of the corporations and communities demonstrate that no one group controlled the future of steam transit development: corporations and communities both attempted to place their will on the other and push for their own advantage. Negotiation among different groups was continuous, and the development of transit was a constant process throughout the antebellum era. Thus, this give-and-take between corporation and community continued throughout the entire period.

Community Impact on Steam Transit

For historians of technology, one particularly fruitful line of inquiry has been to examine how users of a technology change or alter it from what the original designers intended.[81] Such studies throw into question the primacy of the lone inventor in determining the meaning of a technology and highlight how users can play a crucial role in determining a technology's ultimate use. Of course, steam transit could not be brought into the home, which would seem to render this type of interplay between user and technology impossible. But this does not mean that individuals and communities did not attempt to shape the new infrastructure to their own ends. Additionally, users and observers spread steam transit into the broader culture, through their writing and imagery. Such was the widespread nature of this development that by the eve of the Civil War, the awe and wonder that had accompanied the birth of steam transit seemed passé in areas that were saturated by steam transit.

More so than steamboats, railroads were adopted by people for their own ends. Railroad infrastructure created shelters (depots) and clearly marked pathways (tracks) where none may have ever existed before. Despite the danger inherent in using railroad infrastructure beyond its intended purpose, many people found it irresistible to do so, and the challenges for corporations began almost immediately. In 1836, for example, the Boston and Worcester Railroad noted that people were sleeping overnight in their depots.[82] Another challenge to corporations was people walking along railroad tracks. Some reported that walking on tracks could lead to thrilling encounters. A writer in *Ladies' Repository* shared a story of walking along the railroad track in 1849. The writer started crossing a bridge, assuming it would be safe to cross before the train came. Merely halfway across the bridge, however, the author "heard the steam horse ripping and rushing, and tearing and snorting, behind me." With neither time to go forward nor back, the author "leaped from the bridge on to a telegraph post, which stood upright in the water, a few feet from the track, and clung there, like a cat frightened by dogs, until the train had dashed by me. The rush with which the engine passed overset all my notions of velocity."[83] Walking along the track required one to have alert senses: the aural warning of the train's approach might be the last thing heard by a pedestrian.

Other pedestrians did not live to write about their experiences in national magazines. The Boston and Worcester Railroad reported in 1849 that a man walking on the track, "apparently intemperate," was "incurably wounded."[84] In 1860, Andrew Charles Moore wrote that he got injured while walking along the railroad track. It was night, and he could "scarcely see the tracks let alone see the cow pit" which paralleled the track to prevent cattle from entering, and into which he fell and injured himself. Moore hoped he would "not do so foolish a thing again soon."[85] Railroad corporations conceded accidents but also worked to deflect blame. For example, the Northern Railroad (in Massachusetts) acknowledged two accidents in 1849. But the company pinned the blame on the victims: in one case, the person killed was "sitting on the track . . . and believed to be intoxicated," and in the other the victim was "a deranged person, walking on the track after dark."[86] In both instances, the allegations of character and mental disability would be difficult to refute, since the victims were dead, making it difficult to hold the railroad company responsible or accountable. Corporations thus sought to parry blame and even despaired that there was not anything they could do to prevent such deaths. In 1851, the Boston and Worcester Railroad doubted that any law preventing walking on the track would work, since "the highest penalty, death, so often unhappily inflicted, would seem as effectual as any enactment," but the deaths each year did not stop the practice.[87] What made tracks appropriate for trains—flat, clearly marked, free from debris, and well drained—made them equally enticing for pedestrians, despite the inherent risks.

Nevertheless, railroads did work to warn the public at crossings. In 1834, the Boston and Worcester Railroad's board of directors authorized the superintendent to erect signs at public crossings warning about the danger of approaching trains. These signs were to be placed on posts and stretch over the road itself, with lettering "in large & conspicuous letters." But visual cues would not be enough. In addition, the board decreed that each engine should have a bell on it that the engineer would ring before, during, and after passing through a grade crossing on a public highway. Finally, the board asked that gates be constructed at certain public crossings to prevent people from getting on the railroad tracks.[88] Later, the state of Massachusetts enshrined this in law: railroads had to put up warning signs reading, "RAIL ROAD CROSSING—LOOK OUT FOR THE ENGINE WHILE THE BELL RINGS" in letters no less than nine inches tall.[89] When railroads adopted the bell as a warning, they joined a long history of using aural indicators to communicate information. The use of the bell allowed railroads to assimilate aural signaling with which people were already familiar.[90] Other corporations emphasized the sound as well. The handbook for conductors on the Western Railroad noted, "The engine bell will be sounded when the train is within eighty rods of a 'crossing,' (which it will approach slowly,) and continue to be sounded until the crossing is made."[91] The aural signal was to be used precisely to alert all who were approaching the tracks.

Instructions to employees were explicit on this matter, because corporations believed that if they posted signs and rang bells, they were fulfilling their essential duty. In defending the corporation against accidents, officials would point to their posting of signs and use of aural signals as justification that they had done all they could. In 1851, an official on the Boston and Worcester Railroad wrote, "I trust no one questions our legal right to use our track providing we do take the usual precaution of ringing the bell &c."[92] In another incident, the same official disclaimed responsibility for an accident, pointing out, "There was a sign erected at the crossing notifying Mr Bellows [the victim] to 'look out for the Engine when the Bell rings.'"[93] Nevertheless, conditions could make it difficult to heed warnings, as wind and weather could play havoc with the aural cues. In 1856, William Appleton reported that he "came very nigh being run against by the train" because "the rain & wind prevented the man being out with the flag & my hearing the whistle; I was quite near the engine when it passed."[94] Corporations laid out their ground rules for operating safely, but circumstances could still leave the situation dangerous.

By the close of the antebellum era, the concept of walking or sleeping on the track had entered broader cultural conversation. A lengthy temperance broadside excused the railroad in the matter of a death of a drunken man who had fallen asleep on the tracks near Pittsfield, Massachusetts, and was "torn to pieces" by the train. The broadside quoted a speaker who asked who was responsible for the man's death. The speaker did not blame "the ponderous engine, rushing with whirlwind speed over its iron road." Rather, blame fell on "the vendor of intoxicating drink."[95] Another temperance image also blamed alcohol for death on railroad tracks. David Claypoole Johnston captioned a drawing of a man on the tracks in this way: "About to escape from slavery by railroad (not the underground) no uncommon occurrence among slaves." Here, the "slavery" is alcoholism and the enslaved is the alcoholic; this drawing was part of a larger temperance work by Johnston. In the image, the man holds a bottle and is about to experience the deadly force of the onrushing train (fig. 1.1).

As another example of steam transit being used in ways unanticipated by its inventors, steam transit could serve as an attraction for crowds or public

Figure 1.1. Temperance artwork features a "slave" to alcohol about to be run over by an approaching train. David Claypoole Johnston Collection (box 11, folder 4.14). Catalog record 359804. *Courtesy, American Antiquarian Society.*

events. John Bear was a stump speaker for the Whig Party, and he wrote of the reception that he received in Savannah, Georgia. His arriving train was "received with great pomp" several miles outside the city by "over three hundred men." This group then "escorted" the train "into the city to the head quarters of the Young Men's Clay Club, where there were hundreds of men, women and children, white and black, all anxiously waiting to get a glimpse of a live Northern working man who could make a speech."[96] Bear's entrance into Savannah was made all the more triumphant by the procession of the train. The assembled crowd disregarded the train's rules and schedule and made the train's entrance happen on its own terms.

There were other ways in which steam infrastructure was pressed into service by the broader community, showing how the wider culture influenced steam and steam became a more natural part of everyday life. There were circuses that traveled by steamboat and even towed behind them "showboats" to stage their performances. One such showboat was advertised as a "vast Floating Amphitheatre" that could accommodate thousands of spectators.[97] Railroad depots could be prominent meeting places. Whigs of New London

County, Connecticut, for example, listed a series of places where they had met: a courthouse, the railroad depot, a hotel, and the central wharf.[98] An engine house in Syracuse, New York, was the site of a celebration for the first anniversary of the rescue of Jerry, an enslaved man, and fifteen hundred people gathered to celebrate the anniversary of the event.[99]

In 1850, an agent of the Fitchburg Railroad in Massachusetts investigated the possibility of having a concert by famed vocalist Jenny Lind in a hall over a station house. The agent was reassured by the building's architect that "it would be perfectly safe for any number of persons that could obtain entrance."[100] The *National Era* later reported that nine thousand people attended her concert "over the Fitchburg Railroad Depot."[101] The *Jenny Lind Comic Almanac*, published the following year, included stories about the "fever" that gripped wherever Jenny Lind appeared. The book illustrated this fever by making a reference to the railroad in an illustration titled "Engineer of the Atlantic Paying His Respects to Jenny Lind," in which an ardent bearded man gestures outside Jenny Lind's window as she smiles down on him. His idling steam engine stands in the background, presumably blocking traffic in the street and preventing his passengers or goods from reaching their destinations (fig. 1.2).[102]

꩜

When steam transit was new, people had to be educated about what it was. As the antebellum era progressed, steam transit became so well known that writers clearly expected their audiences to be familiar with steamboats and railroads. In 1840, a writer in the *Common School Journal* made a suggestion that demonstrated that he expected the size and shape of steamboat or railroad tickets would be commonly understood by his audience of teachers. The writer used those tickets as the point of comparison when suggesting that teachers could "promot[e] diligence and good order among pupils" by creating series of cards. The appropriate card, reflecting the student's performance, could be given to a student to take home to their parents to communicate how the student was doing in school. The writer urged teachers that "the expense ... would be very inconsiderable, as the tickets need not be larger than those used in steam-boats and on railroads."[103] Such tickets were presumed to be so commonplace that no additional explanation was necessary—the readers would instantly know what size of paper was required.

When antebellum observers looked back on the early period of transportation development, they marveled at how much progress had been made in just a few short years. In 1839, George Templeton Strong speculated what his

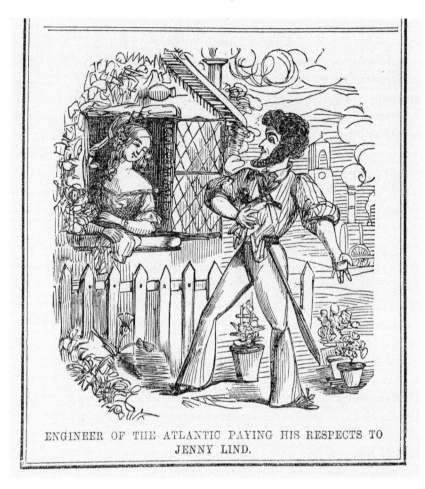

Figure 1.2. A train waits in the background while an engineer gives "his respects to Jenny Lind." *Jenny Lind Comic Almanac, 1851* (New York: Elton & Co., [1850]). Catalog record 196720. *Courtesy, American Antiquarian Society.*

ancestors would think of a train and imagined that they would be astonished by the sight: "whizzing and rattling and panting, with its fiery furnace gleaming in front, its chimney vomiting fiery smoke above, and its long train of cars rushing along behind like the body and tail of a gigantic dragon—or like the d–l [devil] himself—and all darting forward at the rate of twenty miles an hour. Whew!"[104] But steam transit quickly became commonplace. In 1848, the publication *Knickerbocker* noted "RAILROADS have become so common, that the great marvel which they present is almost lost to the general eye."[105] The *Nineteenth Century* concurred that same year: "Though Steamboats have been but forty and Railroads hardly ten years in being, these have ceased to be wonders."[106]

One memoirist noted how quickly railroads had been normalized: "We are now so familiar with their construction, their management, and the running on them, that it is almost impossible at the present day, to realize how little was known of them."[107] In 1852, Caroline Healey Dall wrote in her journal that she had seen a book about the Boston Railroad Jubilee, which "seemed like a breath from home."[108] Once railroads had been strange and different, but for Dall seeing it in a book now provided a welcome connection to home. A newsboy's appeal from 1854 acknowledged that in the future, the advancements of their own age would be seen to be quaint: "And future travellers talk, in pitying strains, / Of sires, confined to creeping railroad trains!"[109] Even animals grew accustomed to the noisy intrusion in their midst: D. J. Barber wrote in his journal of 1859, "Cattles and cranes range through the swamp. The cars will rush along within pistol shot of the latter without causing them to fly."[110] To animals and humans alike, steam transit became normalized.

The process by which this naturalization happened was complex and multifaceted. Initially, railroads needed explanation from boosters. But in just a few short years, the railroad had gone from a talked-of marvel to a commonplace thing or even a pleasant reminder of home. This happened because steam transit did not remain in the exclusive province of economics, whatever the goals of railroad boosters. Through the complex interactions between the corporations that promoted steam transit and the communities that welcomed steam transit, railroads and steamboats had become thoroughly familiar to Americans, and steam transit began moving into other areas of culture. The next chapter shall examine the chief way in which people became comfortable with transit, by looking at the travel experience itself.

CHAPTER TWO

Travel

In 1848, Isaac Hinckley, the superintendent of the Providence and Worcester Railroad, produced a broadside that promised easy connections between his railroad and a steamboat service to Newport, Rhode Island. Passengers could "procure through Tickets" on either end of the trip or on the boat itself. The broadside noted that, to ease movement from the railroad to the steamboat, "Passengers and Baggage transported free of charge, between Cars and Steamboat at Providence." Once in Worcester, further connections were available to towns in New York. This advertisement encapsulates a revolution experienced by antebellum Americans, from travel as a series of difficult decisions about how to get from point A to point B to travel as a commodified service that could be purchased.[1]

Prior to steam transit, travelers had to shoulder much of the responsibility of getting from one place to another. They had to determine their own routes, ask for directions, ford streams, keep luggage balanced on their animals or their own backs, respond to local weather conditions, and meet a host of other challenges. Travelers had to determine for themselves if it was too dangerous or not to move ahead. And they had to secure food and lodging at a time before the widespread availability of restaurants and hotels. There was no "marketplace" to anticipate these types of needs. After the invention of steam transit, however, the entire frame of reference for travel shifted. What travelers could now purchase—in the form of a ticket—was "*travel itself.*"[2] No more did individual travelers have to worry about how to get from one place to another; the railroad and steamboat companies managed the routes. New challenges for travelers would spring up, of course—managing transfers, fending off the grasping agents of competing hotels, and so on. But the fundamental nature of travel had shifted, broadening the number of people who could travel, which in turn led to cultural changes around travel.

One such example of cultural change was the developing standards of behavior on steam transit. As people thronged onto trains and steamboats, they had to determine how to act around crowds and determine how to act when their behavior was on public display. Some of the regulation of behavior was governed by the corporations. For example, time management was crucial to the operation of railroads, and passengers quickly learned that they needed to be on time if they wished to take the train. Other arrangements, such as whether a person could reclaim a seat after vacating it, were regulated instead by the passengers themselves in the moment. Examining the sensory impact that steam transit had on passengers gives us another window into understanding the cultural change that steam transit brought. Transportation brimmed with sensory experiences: the push and thrust of the crowd when boarding a steamboat, the shrill whistle of a departing train, the rushed eating at a station stop, the stench of a boat's cargo—or perhaps a neighboring passenger—and so on. Passengers commented extensively on the different senses as they were activated, and the sights and sounds of steam transit remained with them long after the journey was over. This commentary shows the visceral impact that steam travel could have.

Finally, travelers in the steam era had to assess their own comfort level with the risk inherent with steam transit. Traveling was dangerous even before the application of steam power—a horse could throw its rider, or someone could twist their ankle while walking, or a stagecoach axle could break, stranding the passengers. But steam transit changed the scale and scope of potential accidents. Death on steam transit was a serious departure from the "Good Death" hoped for by observant Christians, meaning that steam transit cut to the core of a cultural expectation around a major milestone in life. Accidents seemingly struck at random, killing one person while sparing their neighbor. And the open design of American steam transit meant that accidents paid no heed to distinctions of class, race, or gender. But as transportation networks grew in the antebellum era, so did the number of passengers. Americans learned to accept a certain amount of risk in steam transit in order to take advantage of its benefits.

Shifting Traveling Perspectives

Accounts of early travel, prior to widespread steam transit, highlight some of the persistent challenges of early travel: the threats of weather, uncertain routes, the difficulty of finding food or lodging, and the level of personal risk that people were willing to endure. In 1793, William Williams wrote about the difficulties of traveling in Massachusetts. Concerned about an impending snowstorm, he decided to set out for home. The snow made travel challenging: by sunset, "the drifts had become too deep for me to travel on foot, and it was with difficulty the horses could get thru' some of them." Williams was concerned, however, that he would get completely snowed in if he did not continue, and thus his group "pushed" forward. But the snow was unrelenting, and eventually the path was "so buried and lost that neither the horses or driver could find it and we were obliged to turn in about seven miles from home." Out with a small group alone against the forces of nature, Williams had to make his own judgments about safety, and if he could not make it to his destination, he could only hope that there was lodging nearby that could take him in. Safely home after the trip had ended, Williams mused that the "journey of life" was parallel to the trip he had just taken through the snowstorm: "Juvenile fire impels us to action; and judgment in this early stage being immature, often without consideration and more frequently without attending to consequences, we plunge into troubles out of which we know not how to extricate ourselves."[3] By drawing this parallel, Williams acknowledged the risks he had incurred by forcing the trip. With the only way home by horse or on foot, these risks were necessary if he was to travel.

Other accounts of travel on horseback, from throughout the antebellum era, illustrate the challenges of terrain and lodging where steam had not yet changed travel. In 1837, Lincoln Clark described a ride on horseback through Alabama. When a bridge had washed out, he was "compelled to strip my nag of saddle and baggage, get into a canoe, and lead my animal by the bridle along-side of our little craft" to cross the "small but angry stream." Later in the journey he secured a meal when he found a country inn "where, though poor, they gave me most excellent dinner of boiled eggs, sweet milk, and fresh butter, I have not eaten so sweet a meal for three months."[4] Safe travel depended on the traveler's ingenuity and locating people along the way who were willing to provide food.

Lodging was also an obstacle. Where taverns or inns were not common, people lodged in private homes.[5] Thus, the quality of accommodation was obviously uneven. When traveling through the Alabama countryside in 1849, Elizabeth Steele Wright commented on the people with whom her group lodged. At one, the landlady served "a tolerable dinner but she had no flour so we had to eat hoe cake without any butter." Wright noted that this quality of food was not uncommon but it "answers very well when a person is hungry." At another stop, breakfast was "first-rate" and a "fine luncheon" was had for "only two dollars. Where we get the best fare there they are moderate with their bills," Wright concluded. But not all food was acceptable. Once she was given "coffee without any cream or sugar," and she noted, "I think if we had to drink coffee that way, that we would not drink much." Breakfast the next day was not much better: "Our breakfast was the same as the supper with the addition of fried sweet potatoes."[6] Traveling by foot or horseback left one at the mercy of local residents for food and lodging, but the unevenness of the provisions and lodging was accepted as the cost of travel given that there were no other options.

Riding in a stagecoach or carriage could hold some advantages over walking or horseback riding, such as the possibility of being protected from the elements. There were, or course, other dangers. Thomas Bull reported on a fifty-five-day journey in 1814, and after one accident "the gig was dashed to pieces and Mr Prentice's ankle sprained so bad that he was unable to walk."[7] Even when the equipment survived the journey, the trip could be physically taxing. Alexander Bliss took a stage journey as part of his honeymoon in 1825. Part of that journey was over a turnpike constructed out of logs. The experience was not always a pleasant one: "In some places earth has been thrown upon them so as to fill the interstices & render the travelling upon it quite easy," he recounted, "but for the greater distance the carriage passes from log to log. This makes the riding extremely hard, painful, & fatiguing."[8] The slow speed of traveling by road meant that sometimes people traveled through the night in order to reach their destinations. Reporting on an 1828 journey, Edmond Kirby wrote, "I got here this morning at five oclock safe & sound except divers bruises from the rough road."[9] Augustus Pleasants Webb departed Baltimore on October 16, 1828, at 3 a.m., and the trip was physically painful. He wrote that he "experienced much fatigue, by being compelled to sit mostly in one position, from the stage literally being crammed, and being unused to travelling and at the same time exceedingly cold."[10] Stage travel created a myriad of possible discomforts, from unpleasant schedules to painful journeys.

When we look at early descriptions of steam travel, it is clear that people immediately saw the advantages of steam over stages, horseback, and walking, even if they did not fully understand the physics of the new modes of travel. People noted their first trips on steam transit in their diaries or in letters to friends, marking them as important events worthy of remembrance. A Mrs. S. Hopkins, who traveled on a steamboat in 1812, wrote in a letter that she completed the journey "less fatigued than I expected" because most of her trip "was rendered very pleasant by the conveyance of the steamboat." Not only was the boat itself "commodious," but the people whom she met on the boat were "of the superior order of beings. Consequently we were favor'd with a little society."[11] Hopkins found steamboat travel to be fast and comfortable, and it allowed mixing with society; it was not cramped like a stagecoach. She also felt that the quality of the fellow passengers was higher.

In 1817, after the inauguration of President James Monroe, newly ex-president James Madison took a steamboat home. The trip allowed the former president to relax, and he could move about the boat. A passenger reported that Madison was as "'playful as a child,' [and] he 'talked and jested with everybody on board.'"[12] Passengers also noted that steamboats allowed them to take in the landscape in a new way. For example, R. B. Mason wrote a letter in 1821 describing his journey on a steamboat on Lake Michigan. He recounted that he "never enjoyed myself more in all my life, than I did, on board of the steam boat." The weather was wonderful and the boat made a stop at Mackinac Island, "which is one of the most beautiful places on approaching it, I ever saw."[13] Before long, passengers complained when they had to return to older modes of travel. Henry Waller was frustrated in 1829 that his steamboat could "go no farther on account of the low stage of the River" and thus he would "be compelled to take the stage at this place ... which will occupy 7 Days and be a disagreeable & expensive route."[14] Steam transit presented a genuine improvement in travel, and Waller was unhappy that he had to use the stage to reach his final destination.

Similar revelations awaited those who took their first ride on the railroad.[15] As with steamboats, early riders of the railroad found those trips to be worthy of inclusion in letters to friends and family, or commented on these first trips in their diaries. Martha Parker could barely contain her excitement in 1834, when she was around nineteen years old. "Have you had a ride on the railroad yet. I have," she bragged to her father. A friend had invited her and some others to inspect a newly constructed railroad. She reported having a "grand time" on the outing. The workers offered to give them a ride on the "mud cars" being used to remove the materials being excavated. "We did not sit down, but rode standing," she wrote, "so you can imagine that we did not go <u>remarkably</u> swift. Not more than <u>20</u> miles an hour."[16] The underlining in her last sentence underscores the irony—twenty miles an hour in 1834 was indeed "remarkably swift" compared with other modes of transit.

That same year, Christopher Baldwin described his first train ride. Seeing the train, he called it "an object of wonder!" and declared that his first ride "seemed like a dream."[17] Likewise, William Elliott recounted his first travel on a train in 1836. As for many other early travelers, the speed of the railroad is what caught his attention. He noted that he traveled at seventeen miles an hour and even, for a portion, twenty-five miles an hour. "The celerity is delightful, almost sublime," he wrote. "To behold a fiery car in front, rushing on, with its long train of chained and captive cars behind, as if bent on some demoniac errand—realises the fabled enchantments of the Genii."[18] In his journal in 1838, Sidney George Fisher reported that a trip of 120 miles on the railroad "seems like magic." Fisher admitted to this feeling of magic even though he had not particularly enjoyed the journey that he had been on, complaining that it "on the whole was unpleasant, from the heat, sparks & crowd of vulgar people." Despite the discomfort, the ability to travel via steam immediately affected his perceptions of other forms of travel. "After steam," he wrote, stage travel "seems a lumbering, slow, tedious way of travelling."[19] For these early travelers, steam transit was not yet naturalized; it was described as magic or a dream.

The paradigm-shifting reality of steam transit was readily apparent to those who were living through it. In addition to comments on speed, people noticed that their perception of time and distance changed.[20] Francis Lieber commented in 1835 that he asked a steamboat captain in Buffalo where he was going. The captain answered, "Chicago," which was some eleven hundred miles distant. Lieber expressed astonishment that such a distance produced nonchalance in the captain: "And this he said in a tone in which a waterman on the Thames would answer a similar question, by 'To Greenwich, sir.'"[21] Perspective of distance had altered so much that a long trip could be treated with seeming indifference. In 1842, Richard Thompson marveled that the trip from Washington, DC, to Baltimore could be done in two hours, rather than one day. He concluded that "there are few objects of more grandeur than a train of rail-road cars."[22] Expressions of awe continued throughout the antebellum era. In 1851, George Smith noted in his journal that he took the railroad and "rode, or rather flew to Indianapolis by half-past 12 o'clock."[23] In 1858, William Elliott was in Saratoga Springs and reflected on the changes in travel during his lifetime: "I came on Saturday by the Hudson river rail road—having accomplished the distance in eight hours—Which, once on a time, required 5 days."[24] That same year, William Floyd left Virginia and reached Memphis,

Tennessee, "in 48 hours. It used to take 28 days on horse back to make the trip."[25] The time savings offered by steam transit were remarkable and did not escape people's notice.

Indeed, by the 1840s steam travel had grown in popularity and prevalence, and it was possible for someone to write that *not* traveling by steam was a completely new experience. For those who lived where their travel needs could be fully met by steam, steam transit could completely dispense with other modes of travel. In 1843, a writer in the *Ladies' Repository* wrote about *not* traveling by steam as "something full of novelty." The writer had experienced only the "certainty of steam navigation." Therefore, "to be dependent on winds and tides, was at once new and strange; and I must confess that a feeling of apprehension, to which I had formerly been a stranger, would now and then flit across my mind."[26] The regularity of steam transit had, for this writer, conquered the uncertainty of nature. Depending on nature was an uncomfortable sensation for those who had never experienced it. What had been normal for centuries suddenly created a "feeling of apprehension" in the mind of the traveler raised in the steam age.

Coming to depend on that regularity, travelers would even delay their trips if it ensured that they could take the railroad. In 1852, Henry Waller wrote that he "stopped all night at Albany, rather than pass the night on a crowded river boat; and came down very rapidly and comfortably this morning on the Railroad."[27] Even people who complained about railroad travel realized that it brought advantages that other forms could not match. George Templeton Strong took a steamboat in 1850 and wrote in his diary that "a decent sloop would be better than that filthy railroad." Yet at the same time he confessed that it was "aggravating . . . to see the trains come squealing and stinking after us as we lounge up the river and rush past and out of sight as if our meek little steamboat were at anchor."[28] Strong may have disdained the dirt of the railroad, but he was still envious of its speed.

Selling Steam Transit

Although historians have often studied steamboats and railroads in isolation from one another, different modes of transit were often closely connected in the travel experience of antebellum Americans. Networks of railroads, steamboats, and stages created a multitude of possibilities for antebellum travelers. The commodification of travel made it easier for travelers to combine these modes: they simply had to purchase the appropriate tickets. To be sure, different types of transit could be in competition, but often companies were cooperating to bring a wider array of destinations available to the traveling public. For example, in 1837, the Philadelphia, Wilmington and Baltimore Railroad printed its schedule on a broadside, making note of the connections it offered, the convenience to travelers, and the scenery along the route. The journey began in a steamboat, which passengers took from Philadelphia to Wilmington. In Wilmington, passengers would switch to the railroad as far as the Susquehanna River, take another steamboat to Havre de Grace, Maryland, and then take another railroad to Baltimore. Once in Baltimore, travelers were promised that they could "take passage South and West without removing your Baggage." The broadside illustrated how multiple modes of transit were integrated with each other to allow people to travel farther. The company would also ensure that baggage made it safely to the destination without the passenger having to lug it from one mode to another.[29] Travel no longer required individual passengers to work out each leg of the trip. One just needed to purchase tickets, and the company took care of the details.

Corporations incentivized interoperability in other ways. This can be illustrated by an example from 1841, in which the Norwich and Worcester Railroad of Massachusetts board of directors voted to give a 12.5 percent discount to any "Stage proprietor" who purchased at least fifty railroad tickets at one time for resale to stage passengers.[30] In another case, a steamboat line in Utica, New York, advertised "A number of Berths and State Rooms reserved for Rail Road passengers."[31] These passengers would not need to fight for space, but their passage from one mode to another would be assured. Likewise, passengers on the Schenectady and Troy Railroad (of New York) had their transfers eased by receiving a ticket from the railroad conductor for their steamboat trip, enabling straightforward passage between the two modes of transit. Additionally, luggage was "conveyed ... free of charge" from one mode to another. The advertisement also noted that the boats would not depart "before 8 P.M. unless the Train of Cars reach there before that time, so that travellers need not fear being left" behind.[32] By easing ticket purchases, promising smooth transfers, timing arrivals and departures, and handling luggage, steam companies worked to give travelers as many potential destinations as possible and in so doing took many of the previous travel hassles out of this new experience. By commodifying travel, they offered real value for the money.

These expanding networks and cooperation among different modes of travel benefited passengers. In 1847, Daniel Helm wrote a long letter that outlined the travel he did while looking for work as a teacher. His roaming was clearly facilitated by multiple modes of steam travel, giving him a wider range of possibilities than previously possible. He traveled from New York to Baltimore through a combination of steamboat and railroad. He then took a train to Fredericktown, Maryland, and taught there for three months. He then took the train to Winchester, Virginia, and taught at nearby New Town for another few months. He then got on the train again at Winchester and traveled to Cumberland, Maryland, and then took the stage and steamboat to Pittsburgh and from there took a steamboat to Cincinnati. After failing to find work either there or in Kentucky, he took another steamboat to New Orleans. When he determined that he "did not like the place," he boarded the steamboat again and returned to Louisville. Steam transit allowed Helm to efficiently traverse the eastern United States in search of work.[33]

When people praised the time advantages of steam transit, it was then to those companies' advantage to ensure that timetables were met. Part of this effort meant trying to shape passenger behavior. Railroads, for example, encouraged people to pay for their tickets prior to boarding. This meant that the conductor would not have to spend time selling tickets on the train. In its advertisements in 1843, for example, the Boston and Worcester Railroad noted that "Fares are less when paid at the Ticket Offices than in the Cars."[34] Several years later, the same company released its rates but also published the following warning: "CONDUCTORS will take 5 cents more than these rates when the fare is paid in the Cars—excepting only when there may have been no Ticket Office open for the sale of Tickets, or when for unavoidable cause they could not be obtained. '*Want of time*' cannot be allowed as an excuse."[35] By charging extra for purchases made on the train, the railroad attempted to enforce timeliness among its passengers. Passengers had to be sure to allow enough time for the purchase of tickets, or would have to pay a penalty.

Indeed, traveling via railroad was governed by time, since time management was the chief means of safety. To keep from crashing into one another, trains had to keep to a schedule. While traveling, travelers had to ensure that they, too, were on time. In 1847, Lincoln Clark lamented that he had just missed his train: "We snatched our baggage and hurried the driver to the depot with all possible haste, but lo! the cars had gone about three minutes before our arrival—the smoke was distinctly in sight." He commented on the reactions of the passengers around him who were also late: "Some grumbled, some swore, and some looked blank."[36] All had been left behind because they had missed the time of the train, and the short time did not matter. Three minutes might as well have been three hours. The commodification of travel brought advantages to the traveler but also meant that travelers had to stick to the corporation's schedule, and learning that lesson could be painful.

As travel transformed into something to be purchased, myriad jobs sprang up to meet the needs of travelers—both real needs and those pushed on travelers. Antebellum travelers frequently commented on the fact that disembarking from a train or steamboat would lead to being set upon by numerous porters pleading for their business—carrying luggage to the hotel that employed the porter. In 1837, Henry Tatham recalled that in Albany, New York, "these villainous reptiles beset us in hundreds, begging and yelling for our baggage." Even when prevented from getting close, porters made the attempt. Tatham recalled that in Utica, "our baggage was deposited in front of all the passengers in the centre of a large enclosure. Not a single porter or vagabond of any description was permitted to come near us. But such screaming and bellowing for our baggage I never could have imagined. We were protected by an open railing through which they put their arms and noses." Eventually Tatham held his umbrella ahead of him as a battering ram and tore headlong into the crowd. But even that did not dissuade the porters; while "some cursed," still "others not at all affronted solicited my patronage."[37]

When William Banister transferred from the railroad to a steamboat in Providence, Rhode Island, in 1837, he found "what constitutes no inconsiderable portion of the miseries of travelling; I mean the scrambling, & pushing, & pulling, & scolding, & almost fighting, for the baggage." Clearly, antebellum porters were not shy about going after business. Banister concluded that porters had no scruples about laying claim to luggage: "One which says 'this is my carriage, sir, this is the one you engaged.' Another says, 'you engaged my carriage, Sir, & I have your trunk,' while a porter has your travelling bag, halfway up the wharf."[38] Indeed, another writer warned travelers that porters would simply pick up bags without warning, "after which they consider you a drawn prize for their particular house."[39] The work of these individual porters demonstrates the expansion of travel's commodification: another task previously done by individuals—carrying luggage—was now a service to be paid for.

The hustle of the railroad depot came in for comic lampooning in a comic almanac from 1859, which printed an illustration titled "A Scene at the Hudson River Railroad Depot. A Countryman Taken by Surprise on his First Visit to the Empire City." A man flops over his luggage shaking a fist at all who are trying to get his business by carrying the luggage, while getting his pocket picked by a youth. Packages are being thrown from the top of the train down to the people on the platform below. Those looking on are laughing at the man's naïveté and misery. The image shows the chaos of the station as well as the futile attempt of the individual traveler to be left in peace while trying to get to his destination (fig. 2.1).

Above and beyond the challenges of porters, passengers had other reasons to pay close attention to their own luggage, for fear of it getting lost. In 1849,

Figure 2.1. A first-time traveler struggles to hold onto his luggage, to the amusement of others on the platform. *Rip Snorter Comic Almanac 1859* (New York: Philip J. Cozans, [1858]). *The Library Company of Philadelphia.*

Horn's Railroad Gazette warned passengers to always ensure that they got a check from the proper baggage master, and not fall victim to "officious persons . . . who with great apparent frankness offer you checks for your baggage." The newspaper asked a blunt question: "Are you certain that you did not receive a spurious check, and that your trunks instead of being deposited in the baggage-car were not conveyed away secretly?"[40] This article exploited the fear that a criminal could make off with an unsuspecting traveler's luggage. Luggage could also get misplaced by accident, despite the best efforts of passengers. While we will probably never know the exact scale of the problem, a list of unclaimed items published by the Western Railroad Association in Chicago in 1854 suggests that a wide range of materials was left behind on trains. The list includes some unusual items: "stone cutters' or mason's tools," "egg of maple sugar," "dirty white shirt," "4 vols. of Alison's History of Europe," and "free papers of slave Eliza."[41] Clearly, all types of curious—and vital—things were left aboard transportation.

When bags were lost, passengers had to go to great lengths to retrieve them. Henry Waller's father once left behind a leather portfolio, "containing a large amount of money," which his father did not realize until they were "7 miles on our way" on a train. Seeing his father's distress, Waller left the train "and would have walked the 7 miles back," but fortunately he came across the train going in the opposite direction. Thus he was able to return to his starting point quickly, locate the portfolio, and rejoin his father on the same day. In another case of lost baggage, Waller was more annoyed with his relatives. While at the train station, he left them in charge of baggage and went to pick up a pair of boots he had purchased. When he boarded the train and asked if they had seen the baggage, they responded that they knew nothing about it. He was angered by their lack of worldliness: "They really thought that the car they were in was the only one going to Phila[delphia]. They hadn't enquired a word about it." Waller rushed off the train, had the baggage put in the correct baggage car, and reboarded. As an experienced traveler, Waller picked up specialized knowledge, and he was frustrated with those who were not schooled in the methods of traveling. He described his relatives as "wild & ignorant as if they had just been imported from New Zealand" when he discovered that they had not properly cared for the baggage.[42]

If the world outside a train or steamboat exposed passengers to the challenges of porters and lost luggage, while passengers were in transit, there was consistent authority. Conductors held particular power on a train, as did the captain on a steamboat. Both were in a position to enforce the rules as they saw fit. In 1839, a writer in the *Colored American* complained about a steamboat captain who delayed the trip in order to pick up more passengers from a stage arriving the next day: "Steamboat captains here consult their own interest, and very little passengers haste."[43] In 1850, when Moses Hoge did not have enough money in his pockets to pay the fare, the conductor marched him back to the baggage car and Hoge retrieved his trunk, "where my fast diminishing & nearly exhausted stock of gold was deposited."[44] Conductors and captains could be capricious, but they held the acknowledged authority on their respective modes of transit. Corporations recognized that conductors held a special sway over passengers, and thus the corporations told their employees to use their position wisely. In 1850, the Michigan Central Railroad printed its rules for employees. Passenger conductors were instructed to do the following: "It is one of the special duties of a Conductor to do anything (not inconsistent with other duties, or the regulations of the road) to accommodate passengers; to answer in a proper and civil manner all questions, and endeavor to leave a good impression on every one." Further, it was noted that "from their position they are able to exercise a material influence in turning the patronage of passengers to certain hotels along the line, as well as at the ends of the road. The impropriety of this they will readily see."[45] Conductors had to be above reproof.

Negotiating Comportment

Finn's Comic Almanac of 1835 published a satirical verse that hinted at a broader truth: "Aristocrats sincerely wish,/All Rail-roads at the *d—l;* [devil]/Because the purse-proud and the poor,/Are both brought to a *level.*"[46] Indeed, steamboats and railroads brought together hundreds of strangers for a short period of time to experience travel together, cutting across class, gender, and racial lines. Previous modes of travel were not geared toward large groups: A handful of people might ride together in a stagecoach; walking and horseback riding were even more solitary. As Americans of different classes and life experiences were mixed together in steam transit, they had to learn how to act around one another. The antebellum period, therefore, saw the development of social norms for traveling with large groups of strangers.

Contemporary observers certainly noticed the crowding on steam transit. In 1815, Harriott Pinckney Horry wrote that a "monstrous crowd" was aboard the steamboat *Car of Neptune*.[47] Anna Calhoun Clemson wrote in 1836, "I believe I have met everybody I ever knew since I started the people absolutely travel in <u>droves</u> for no other word will express the masses of human beings one meets in the steamboats and rail road."[48] Likewise, an early cartoon mocked passengers being packed onto a steamboat. With the ironic title "Elegant Accommodations," the image shows a commotion of horses, pigs, people, and carts, with bodies jammed up against one another. At the rear, a man shouts out, "That's it Paddys, drive the passengers into the Hog pens and make room for the milk carts." A woman calls out in distress, "Oh look my poor dear Husband is under the Horses belly." The image title, deriding the promise of comfort offered by steam transit, contrasts with the image itself, suggesting a jumble of humans and animals all trying to hold on to their small patch of deck for themselves. It illustrated the lack of control that individual passengers had over their circumstances; one could not control what other passengers were on a ship or what they brought with them (fig. 2.2). Such scenes highlighted the openness of transit in the United States and the need for norms to govern how people would interact with one another.

Figure 2.2. Passengers, freight, and animals compete for room on a steamboat, highlighting the irony of the title. *The Library Company of Philadelphia.*

Passenger travel in the United States was generally more physically open than that in Europe. Rather than having separate compartments for small groups of passengers, the railroad car had a single aisle down the middle with rows of seats on either side. The result of this physical openness was social openness as well; as historian Eugene Alvarez has noted, the presence of the center aisle "discouraged class distinctions."[49] Naturally, there was a class distinction in that people had to be able to afford a ticket to board. And there were some efforts to segregate passengers, as we will see in chapters 5 and 6. There were other efforts to create separate classes. In 1835, for example, the Boston and Worcester Railroad voted to establish special cars for "labouring people & for such persons as complain of their inability to pay full price."[50] On steamboats, "steamboat owners promoted the idea that cabin passage was a refined experience," making it possible for passengers to purchase their way into a more comfortable way of travel.[51] But establishing these different levels of service could be expensive. On a railroad, for instance, it would require additional cars or for existing cars to be subdivided. Thus, the complete segregation of passengers by class on steam transit was incomplete in the antebellum era.

Most Americans seem to have accepted the mixing brought about by transit. Etiquette guides gave instruction for how people should act in an environment surrounded by strangers and where they were constantly on display, a situation they may not have found themselves in prior to steam travel. One writer, attempting to defend the character of the people of Kentucky in 1828, noted that the character of its citizens was "on the whole estimable" and that the writer drew this conclusion from "an impartial examination of its citizens, in steam boats, in taverns, in stages," and in other places. Steamboats, taverns, and stages, all associated with travel, were places of public display where one could judge the character of a people.[52]

The variegated company was frequently commented on by passengers, which is how we know that not everyone approved.[53] Traveling in the Midwest in 1853, Elisabeth Koren was disappointed with the lack of social separation on the train: "The coaches were filled with an unpleasant mixed company, which one must put up with here where there is only one class." She left Chicago in a steamer, and was impressed by the quality of the boat: "The upper saloon, where we are—there are two decks—is extremely large, lacquered white with richly gilded borders and carvings, four large, handsome chandeliers, several gilded mirrors. The floor is covered with expensive carpets, and there are mahogany tables with marble tops, velvet-upholstered sofas and chairs." But again she despaired over the lack of segregation on the steamboat: "The passengers, however, are not at all suited to this elegance—a mixed company here as everywhere."[54] Steam travel provided a new environment in which Americans had to reach agreement on their interaction with each other; some welcomed this opportunity, whereas others, as *Finn's Comic Almanac* knew, decidedly did not.

In an effort to bridge the gap and make travel more pleasing for everyone, etiquette guides set out suggestions for behavior on transit. An etiquette guide for men published in 1857, for example, explained through a series of vignettes the proper way to act while traveling. In one, a young man, chastised by his father for speaking too loud on a steamboat, bowed to the group of men he had bothered and said, "I beg your pardon! I really was not aware of being so rude."[55] In this way, etiquette guides hoped to introduce social norms to the uninitiated or those new to travel. A person unaware of how to act could read such a guide in the privacy of their own home and learn the right way to behave. But there were still areas where passengers needed to work out their own arrangements in the moment, such as determining rights to a vacated seat.

When seats were not reserved, the taking of an empty seat could lead to public confrontation. In 1848, Frederick Douglass recounted a lengthy verbal exchange between "Out of Seat" and "In the Seat." Out of Seat demanded his place back, but In the Seat refused to budge. In the Seat appealed to an understood rule among passengers: "When a seat is vacant it is as free for one gentleman as another." Out of Seat asked the man in the neighboring seat to uphold his claim, but that passenger remained "silent, evidently wishing to avoid a row." In the Seat pressed his claim further: "A man's right to a seat ceases when he leaves it and there is nothing of his left in it to show that he intends to return to it again. This seat was entirely vacant when I took it." Out of Seat then threatened physical violence. At this point the other passengers joined in on the side of In the Seat, laughing and letting out "cries of Hold to your seat." Out of Seat relented and went "peaceably into another car amid the jeers of the surrounding passengers."[56] In this story, we see how passengers themselves, in the absence of rules from the railroad, laid out their own standards for how to determine that a seat was claimed. The "jeers of the surrounding passengers" and the disinclination of the neighboring passenger to get involved show how the standards were enforced through the actions of fellow passengers. Rather than citing an etiquette book, In the Seat pointed to a rule as it was commonly understood, and he had his rights maintained through the pressure of the other passengers taunting his opponent.

Once seated, passengers then had to determine how to pass the time when traveling. Plenty of passengers commented on the passing scenery as it floated by their window on a steamboat or whizzed by on the train. But there were other means of entertainment. Augustus Pleasants Webb noted in 1828 that on board a steamboat there was an "Italian juggler" who "amused" him.[57] In 1858, a newspaper told of a chess board that promised, "It can be played in any carriage or car, over the roughest of roads, for there can be no displacement of the chessmen."[58] Fellow passengers could also be entertainment. In 1851, Caroline Barrett White reported that she "enjoyed the ride very much," because there was "an interesting love scene enacted before us all the way, by a young couple who seemed to have a wonderful attraction for each other, and the frequent collision of whose lips attested strongly the warmth of their feelings within. Harmless collisions! What destruction if it had been a collision of Engines instead!!"[59] White's use of a train metaphor here underlines the difference between the voluntary and involuntary "collisions": whereas the voluntary kisses held great appeal, an involuntary collision of trains could only lead to horror.

Far and away, however, the most common activity to pass the time was reading: an attempt to create solitude and separation apart from the din of the rest of the passengers on board.[60] Some antebellum Americans worried that the frenetic pace of the age was ill-suited to reflection.[61] But the railroad "provided a new opportunity for reading," not just because trains carried voluminous new reading material to the areas that they served but because publishers designed different types of books just for reading on the train.[62] As historian Kevin Hayes has argued, the anonymity of transit allowed for a wider range of reading: "The latest romance or sensational novel which someone had secretly wanted to read but had avoided at home could be read comfortably among strangers on a train."[63] People could bring their own reading material on the train, and newsboys sold reading material to those who did not bring it. In 1854, the Philadelphia, Wilmington and Baltimore Railroad instructed conductors that they "must not permit the sale of Books, Papers, or Refreshments, in the Cars by any person not duly authorised by the General Superintendent."[64] Corporations worried that people would be overwhelmed by salesboys, or perhaps they did not want competition for their own vendors.

In time, publishers recognized that they needed to make materials purpose-made for consuming while traveling. In 1852, the *National Era* praised *Chambers's Pocket Miscellany* as "seemingly made on purpose to relieve the monotony of railroad traveling; and the whole is put together in a shape having direct reference to that comprehensive nook, the outside pocket of your travelling wrapper."[65] The publication was apparently designed to be tucked into traveling attire and easily accessible while in transit. Highlighting the degree to which book production responded to the desire of reading travelers, book reviews would note which books had particular utility for readers when they were traveling. In 1855, for example, a reviewer wrote that the book under review was "a capital book for a leisure hour or railroad travel, or for those seasons when you want to be pleased without effort."[66] Another book was praised as "such a book as we like to take a long with us in the steamer or rail-car, when on a visit to a watering-place or a summer excursion."[67] A review of the *Knickerbocker* magazine in 1858 complimented it as "one of the best of the magazines, especially to catch up for a few moments, and lay down again, to resume at a leisure hour, or when on a steamboat or railroad journey."[68] Reviewers recognized steam transit of having particular demands, and noted when reading material was appropriate for a trip.

Publishers began developing books explicitly for the railroad trade. G. P. Putnam and Company began "Putnam's Railway Classics" in 1857.[69] Advertisements for the "American Railway Library" and "Bradley's Railroad Library" indicate that other publishers put out series of books with the intent that they would be suitable for reading on the train.[70] Bookstalls appeared in railroad stations, and boys were tapped to sell wares in the trains themselves. Hayes credits "the unique design of the American railway car" for allowing the newsboy trade to flourish.[71] Designed with a single aisle down the middle, rather than carved up into compartments, cars allowed newsboys to easily walk down the aisle and sell goods to people sitting on either side.

If reading was not of interest, travelers could always talk to those around them. Sometimes, transit was a way to meet and converse with old friends. In 1852, Lincoln Clark described his delight in meeting "our blessed friend Mrs Ellis" at the railroad depot, who was "glad to see me."[72] Others looked forward to talking to their temporary neighbors, even if they were strangers, and were disappointed when they could not find a conversation partner. An anonymous traveler complained in 1839, "Few in Carrs & par consequence very dull—no one to talk to."[73] Many antebellum Americans noted the propensity of passengers to strike up conversations on steamboats and railroads. In 1851, Jane Caroline North wrote that during a journey she "met a curious person in Richmond" at a hotel and then when she later boarded the train the same person "took a seat beside me & talked incessantly, bantering me on being very sentimental & a great deal equally entertaining." Although North "wished she would go to her own seat & let me alone," the woman continued for a while, finally exclaiming, "Dont you want to know who I am? you must think it funny a stranger should be as sociable," and revealed herself to be a family friend.[74]

For some passengers, these conversations were welcomed and helped to pass the time. In 1854, Carolina Seabury recorded how people start talking to each other on the train: "The first feeling of being utter strangers to each other soon passed away. As one thing after another called forth little expressions from each of us—and night soon came which seems to equalize all travelers' feelings wonderfully—We had every requisite for a pleasant journey—pleasant company, faultless weather & fast traveling."[75] Traveling allowed for a certain degree of anonymity. People were briefly thrown together for a few hours, intermingled, and then departed, perhaps never to see one another again. As historian Robert Gudmestad has argued in the context of steamboats, the "anonymity of travel on western waters gave men the opportunity to transcend their mundane lives and take on new identities as they crossed into a world of play."[76] Of course, this could apply to railroad as well as steamboat travel. For example, in 1843 Dexter Russel Wright took advantage of this anonymity to amuse himself. He struck up a conversation with a fellow passenger on a train, took an assumed name, and used this subterfuge to ask the unassuming passenger about "a man by the name Wright" (that is, himself) and Wright's fiancée.[77] The anonymity of steam transit here provided Wright some fun.

Conversations could be on any topic, and people did not shy away from controversial questions. This was appealing for some and dreadful for others. In 1844, Thomas John Young complained that politics had "been the theme of Rail Car, Canal & steam boat, bar & drawing room discussion wherever I have been."[78] In 1848, the *National Era* reported that straw polls were being held on transportation in advance of that year's election: "In all the votes taken on board steam or canal boats, or railroads, Cass is universally behind Van Buren."[79] Perhaps most strikingly, numerous commentators, particularly in the Black press, noted the discussion of slavery on board steamboats and railroads.[80] Indeed, a writer in the *Colored American* in 1837 desired that talk of abolition would fill "our coaches, rail road cars, and steamboats, and hotels, and private families, and public meetings, and churches." The writer hoped that abolition could become a genuine possibility if it was so widely discussed. A letter writer that same year recounted his own experience and concluded that a good measure of "the onward march of the principles of Abolition" was the fact that it had become suitable "steamboat and stagecoach conversation."[81] These writers hoped that abolition could be pushed over the invisible line that separated acceptable from unacceptable public conversation.

Numerous reports from the antebellum era attest that this line was breached, and abolition was a frequent topic of public conversation among travelers. A writer in 1838 found abolition "agitated in the steamboat, the railroad car, and the fashionable hotel." The following year, the *Colored American* reprinted an exchange between Quaker John Greenleaf Whittier and a slaveholder in a railroad car, which was "a free interchange of views on the agitating subject, in which the Southern frankly acknowledged slavery to be a shame and disgrace." Another letter to the editor noted that a debate over slavery on a boat was so intently felt that "one of the gentlemen who took part passed by Cattskill, where he intended to land, and landed at Rhinebeck, so interested was he."[82] The man was so swept up in the debate that he forewent his stop.

In 1848, Frederick Douglass reported that he was part of a conversation regarding slavery on board a steamboat on Lake Erie. As he recounted it, when some passengers aboard learned who he was, "they insist[ed] upon me to give them an Anti-slavery speech." The question was put to a vote, and the passengers then voted unanimously to hear him, with even the captain voting in favor. After Douglass's talk, a slaveholder on board then arose to defend the practice. Douglass debated him further, which "led to an interesting discussion in which a number of the passengers participated." Douglass felt that the entire discussion "showed the great change going on in the public mind on the subject of slavery."[83] This instance highlights how steam travel, with its captive audience, could serve as a forum for debating controversial issues.

Other examples come from people less famous than Douglass. In 1849, the *National Era* published a letter from a correspondent who wrote about Kentucky: "I have recently visited Frankfort and Lexington. In the steamboat, the railroad car, and the tavern, *Emancipation* was THE question discussed."[84] And in 1853, a writer recounted that he saw a "colored man, well dressed and of an intelligent countenance," standing at a depot. A stranger remarked to the writer: "A smart cuffee that! I suppose the rogue has run away from some honest man at the South." The writer responded, "It is more probable that he is an honest man who has run away from some Southern rogue." The stranger "instantly arose and went into another car." But the stranger then returned and engaged the author in conversation, during which the author claimed he was able to correct the stranger's "erroneous belief respecting the principles and measure of the abolitionist," and the two men exchanged contact information when they parted. The writer believed that this experience illustrated the proverb "A soft answer turneth away wrath."[85]

In the Black press, such stories of public discussions and debates on slavery were regularly printed. This is striking, since it was not a given that there could be public discussions about slavery in the antebellum United States. In 1835, Postmaster General Amos Kendall instructed northern postmasters that "they were 'justified' if they refused to dispatch abolitionist mailings into the South." Additionally, a "gag rule" prevailed in Congress that prevented the hearing of abolitionist petitions.[86] By contrast, discussion of such topics appears to have been admissible on steam transit. Surely part of this freedom came from the anonymity of transit—people might have been bolder in raising topics with their seatmates knowing that they were unlikely to ever meet again. The design of steam transit may have also played a role. With such broad mixing in the train cars or steamboat salons, it would be impossible to know who your seatmate might be, and so an abolitionist and a slaveholder could well be sitting side by side.[87] Perhaps the freedom of knowing that any such acquaintance was temporary made it easier to enter into a discussion of difficult subjects. Although prohibited from the mails and in the halls of Congress, the topic of slavery and abolition appears to have been hotly discussed on steam transit. While the abolitionist press included stories of successful conversions, it is easy to imagine that not all discussions of the topic ended amicably. The open design of the cars and social mixing may have encouraged risk-taking in conversation, but there was no guarantee as to the outcome.

A Journey's Sensory Impact

Steam travel was an assault on all of the senses, and many travel accounts are replete with sensory description.[88] When traveling, people placed themselves entirely inside the new technological marvel. Thus, there were abundant new opportunities to see, hear, smell, taste, and feel. Antebellum Americans' accounts richly describe their sensory experiences. In 1829, Sophie du Pont wrote that her trip on the steamboat *Albany* was a riot for the senses: "Then commenced a scene, which can neither be imagined nor described, but must have been <u>seen, felt,</u> and <u>smelt</u>, ere one can form an idea of its discomfort & singularity." She described the mass of people on the boat as "moving chaos." She noted the different ways in which her senses were aroused. First came the tactile sensations: she described the press of all the people on board, as "as many were stowed in the cabin as it could <u>inconveniently</u> contain." Next, the sounds: "The laughing of the gentlemen, the groans of the ladies, the squalling of the babies, the rattling of teacups, plates and knives, assailed the ear with a hideous din." Finally, the smells: their noses were "no less molested by the perfume of salt fish, peppermint, fried ham, and other nameless odours as various as disgusting."[89] The experience of travel heightened the senses, as reflected in antebellum Americans' description of their travel.

As we turn our attention to the individual senses, we find that sight received a great deal of comment. Steam transit—with its higher rate of speed—placed passengers into a new relationship with the landscape around them. Passengers often commented on what they could—or could not—see. In 1855, David Auguste Burr's railroad journey offered some splendid sights. Burr awoke on the train while it was in the "Allegheny Mountains, in the midst of some of the most grand & sublime scenery." Burr noted how the trip rapidly offered different visual sensations: "Now we would be flying like lightening speed along the steep side of a high mountain with a precipice on one side & on the other nothing to be seen but rocks & gravel flitting by so rapidly as to make one dizzy. Now passing into darkness of night when we flew through the bowels of some high hill or mt. or else viewing with admiration the works of man when we would pass over seemingly a few cords suspended over a yawning abyss."[90] The experience could be wholly disorienting. Alternatively, travel could provide sweeping views that captivated passengers. Caroline Barrett White reflected on a journey in 1849 that it was good to see "beautiful meadows, surrounded by lofty hills, and covered with noble trees, which it is a real feast to gaze upon after passing through the swamps and hedges which mark the lower end of the route."[91]

Passengers sought out views and were disappointed when they could not see them. Taking a steamboat into New York City, Charlotte Hixon wrote in 1849 that she was "hoping to enjoy a view of the city as we approached, and rose early for that purpose. But a dense fog concealed it from observation."[92] Writing from the Catskills in 1833, Mary Lyon noted that the passengers were truly struck by the scenery as they went up the Hudson River in a steamboat. She noted that they were surrounded by beauty: "On one side, we behold neighboring summits, rising high above our heads; on another, we cast our eye downwards on the waving tops of the thick trees, with their rolling surface; while in a third direction, rival mountains seem to unite, tipped with the blue summits of more distant and higher peaks." Everywhere a passenger looked, the scenery offered new delights. Lyon noted that the passengers were completely distracted by the views: "Scarcely looking after their baggage, almost all, with one consent, stopped to admire and wonder." When passengers went out onto the decks to admire the view, "the low voice and sweet stillness reminded me of a company in a gallery of paintings."[93] Keeping close attention to baggage was critical during the journey, but the visions offered on this route overwhelmed the senses and changed passengers' priorities. Passengers lost thought of their belongings and adopted an attitude more appropriate to a museum.

Travelers also noted the sounds of transportation. Joseph Dowding Bass Eaton wrote that when he took a steamboat journey, "The boat machinery is continually thundering in my ears—a great objection to this mode of travelling, although superior to any other." He noted the following day that in the boat he was "situated as easy as if in a drawing room, were it not for the noise of the machinery."[94] The noise of the steamboat was bothersome, but Eaton still believed that it was preferable to other modes of travel, based on the level of comfort, which reminded him of a relaxing seating area at home. Anna Kauffman complained about the sound of trains coming into her town in 1850: "There is a wonderful screaming here when the cars come in we can hear them long before they get in."[95] The sound of travel expanded beyond just the machinery itself and included the employees. Henry Tatham described the songs of enslaved people who worked on steamships in St. Louis in 1837: "The negro slaves, firemen, were singing their songs in full chorus as is usual. Their deep mellow voices have really a good effect. Their songs are original for instance thus[:] The Bon'slick is a Bully boat / And we're a bully crew / Our captain is a bully / And our clerk's a bully too."[96] The enslaved men sang as they worked, slyly playing on the alternative meanings of "bully" as an adjective (meaning excellent) and noun (meaning a cruel person). The sounds that shaped the travelers' experience were both new (such as the machinery) and familiar (such as the enslaved people singing "as is usual").

The sounds associated with steam travel could also be manipulated. In 1848, the *New Hampshire Gazette* published this story in which a steamboat captain tried to use his whistle to drown out the voice of a politician he opposed. When Senator William Allen of Ohio raised his voice, "so did the noise of the steam-pipe, and thus they had it, whistle steamboat, scream Allen: but the stentorian lungs of the latter prevailed, and far above the noise of the engine was heard the trumpet-toned voice of the tall senator. The captain gave it up—swearing that it was the first time his boat was beat."[97] A tourist guidebook acknowledged that the sounds on board a railroad were many but should not distract from the visual grandeur outside: "We pray the traveller not to let the political discussion among his masculine friends, nor the chit chat with his lady companions, nor yet the newspapers containing the fullest account of the railroad tragedy, or the new book fresh and damp from the press, detract his attention from the scenes without."[98] Steam travel produced and engendered all sorts of noises, and this guidebook implored passengers not to engage with the noise, but to stay focused on the sights.

The sense of taste also elicited comment, as eating—however rushed—was also part of travel, particularly on longer journeys. Passengers were accosted at railroad stops by people selling food. In an 1836 railroad journey, William Elliott noted that "when the train stopped a few minutes at a village—little girls rushed forward to sell fruits to the passengers, large bunches of ripe cherries, and paper cups filled with Gooseberries—(strawberries likewise are in full season)."[99] An article in the *Richmond Whig* alluded to "the old negro women who usually frequent the cars on arrival at Gordonsville, for the disposal of refreshments."[100] In the absence of a dining car, food bought through the window by passengers at depot stops provided fresh and welcomed sustenance during a long journey. Like the newsboys or booksellers previously discussed, the women and girls who hawked food at depots were entrepreneurial in filling the voids left by the railroad corporation, addressing the needs of passengers, and further contributing to the commodification of the travel experience.

Meals were sometimes held at the mercy of hotel proprietors. Samuel Chamberlain recounted that one such hotel owner attempted to cheat passengers out of a meal in 1844 by taking a long time to prepare the food during a layover when there were only twenty minutes between their arrival and next departure. Another passenger sensed that they were about to be cheated and "cried out 'help yourselves gentleman.'" Chamberlain reported that he "obeyed orders with a hearty good will by sequestering a roast chicken and a apple Pie. In spite of the remonstrance of the out flanke'd Landlord, the table was relieved of all its eatables, and we ate our dinner in the cars with our lady passengers to grace the feast, and a right jolly time we had of it. I shared my plunder [with] two elderly Laides, who contributed to the repast by producing a well filled lunch Box and a pocket companion of good brandy."[101] Chamberlain was able to secure his meal—and a good story—and share it with fellow passengers upon return.

A final way in which taste made it into steam travel was that astonishment at the speed of travel was also expressed in terms of food and dining times. In 1833, "An Old Engineer" wrote about a theoretical trip in a newspaper and tied the different stops to different meals during the day: "To Athens [Georgia] to breakfast (on wild turkey,) to dine at Mobile (on oysters or turtle soup,) and to sup at New Orleans (on shrimps) oh it is like sleighing that we hear the yankees talk about. It is like hanging to the tail of a comet." Distance here was marked conceptually by mealtimes, and the concept was picked up by others.[102] For example, writing of Portland, Maine, a guidebook from 1851 noted, "The entire journey from Boston to the mountains may easily be performed in a day. The traveller may eat breakfast in Boston, dine in Portland, and sip his tea in the very neighborhood of this noble mountain."[103] What seemed incredible—eating successive meals in wholly different cities—quickly became a reality of which steam travelers could avail themselves.

Like the other senses discussed here, the sense of touch was omnipresent: verbs of physical dislocation and movement permeate descriptions of steam travel. In 1837, Mary Telfair did not enjoy steam travel in part because of the physical discomforts it wrought. She wrote that after a journey in a steamboat with four hundred other people, she felt that the *"evils of travelling greatly encreased"* by steamboats. She longed for earlier forms of travel, "when people depended upon sloops & stages for transportation they were not likely to be jostled out of their existence."[104] By contrast, James Davidson recorded in his diary in 1836 that he enjoyed sleeping on a steamboat: "The rocking of the Boat and the lumbering of the machinery lulls me to rest."[105] But most others commented on how steam transit was jarring. In an 1843 letter, Lincoln Clark explained that he "could not write on board a boat, there is so much jarring of the machinery."[106] Edmund Kirby wrote that the steamboat he was on "jumps like a dancing girl."[107] Isaac van Bibber wrote of a Maryland journey by train in 1844 that he "read and shook and grunted until I arrived at Sykesville."[108] In 1848, Joseph Mersman found the movement of the train to be first troubling and then more soothing: "During the first night I had some trouble to get asleep, the continual shaking and noise of the Steam rendered every attempt abortive, but during the succeeding one's it seemed to Contribute if any thing to sound snoozing."[109] In 1856, William Chauncy Langdon "took the Galena cars & was jolted and bounced over a most villainous road."[110] Henry Waller noted that same year, "The weather was cold, and quite a storm during the latter part of the trip; so that we nearly froze when off from the fire, and melted when by the little red-hot stove. This process of freezing & thawing kept up through an entire night."[111]

Night travel was a time when comfort was particularly important as people tried to sleep. The Memphis and Charleston Railroad reported in 1860 that it had added "sleeping cars to our night trains" which "proved to be a popular measure, and are eagerly sought for by appreciative travelers."[112] But despite such efforts, passengers continued to report discomfort. Mollie Dorsey Sanford recorded after a trip that she felt "as if I had been shaken up and tumbled out of a four-story window."[113] And William Dutt apologized for cutting a letter short, "not because I have no more to say, but because I can't write, and I doubt whether you can read this, but I can't help it I could do no better this time, so adieu till I write again."[114] Antebellum travelers certainly appreciated steam transit's speed and convenience, but they did not hesitate to point out how it jostled their bodies.

Steam transit, then, evoked for passengers a complete sensory experience. These sensory descriptions give us a fuller picture of how antebellum Americans encountered the machine. For many Americans, steam transit was the first time that they put themselves completely in the hands of a new technology. As they thought about and discussed their experience, they realized that it impacted them in a variety of ways: the way that their body moved (or was pushed!) within a rail car or steamboat, the sounds of the engine or the laborers, and the taste of the food consumed on board. This was not merely the experience of artists and writers, but regular travelers made these observations as well. Sight, touch, taste, smell, and sound all contributed to people's passage on the railroad or steamboat and shaped whether that experience was positive or negative.

Inherent Risk

Any time someone boarded steam transit, they assumed some risk. Danger was, of course, present for travelers before steam: a pedestrian could freeze if they ventured into the snow, or a horse could frighten and throw its rider, or a stagecoach axle could break, injuring all concerned. But steam changed the scale and locus of responsibility of accidents. As Wolfgang Schivelbusch has argued, while a broken carriage axle inconvenienced those in the stagecoach, a broken axle on a train could endanger hundreds. Individuals had individual control over their pace of walking or the reins of their horse, but with steam hundreds in a single boat or train turned over their safety to the workers of the corporation. Finally, steam transit's openness to all meant that women, men, and children, rich and poor alike, could fall victim. Steam accidents were indifferent to social class, age, gender, race, or other distinctions.[115]

The nature of death by steam presented a major challenge to many prevailing cultural attitudes toward death in the antebellum era. As Drew Gilpin Faust has written, the "Good Death" was "central to mid-nineteenth-century America," with Americans understanding that such a death included specific "rules of conduct" on "how to meet the devil's temptations of unbelief, despair, impatience, and worldly attachment; how to pattern one's dying on that of Christ; how to pray." Death by steam transit could be sudden and without warning. Its victims would not have time to adequately prepare themselves for the structures dictated by the Good Death. Almost by definition, dying on steam transit meant dying away from home, not surrounded by family, and unable to communicate any last words to loved ones. As we will see in chapter 4, Americans turned to their religious leaders for help to understand how to reconcile the desire for Good Death with the reality of modern life. Ministers addressed these concerns head-on, imploring their flocks to live their lives as if death could come at any moment.[116]

As historian Mark Aldrich has noted, many different actors contributed to the assessment of risk on railroads, and the same applied to steamboats as well. The corporations who ran these modes of transportation certainly had their own vision of what constituted danger and who was responsible. The press reported on accidents near and far, which in turn shaped public perception. Public outcry could lead state governments to respond by setting down rules and regulations for how corporations had to act. The interpretation of these regulations was then argued out in the courts, giving yet another body a say in what risks the community would consider acceptable. At the same time, despite fears of dangers, it is clear that Americans never abandoned steam transit altogether; indeed, demand and construction only grew throughout the antebellum era. As Aldrich has shown, the earliest decades of railroad development had "many casualties and small-scale accidents, but few major train wrecks."[117] Therefore, while it is important to understand the threat of accidents and to recognize people's concerns, the overall trajectory was that Americans became "inured to the hazards of travel"—or were, at a minimum, willing to accept those risks.[118]

There is no doubt that Americans were aware of the risks they were taking. As *Dinsmore's Road and Steam Navigation Guide* of 1859 pointed out, when on a railroad trip, "You are rushing onward, and you are powerless; that is all." Riders had to surrender themselves completely to the experience. At the end of the trip, the guide noted that a traveler might "draw a long breath, as you dismount at last, a hundred miles away, as if you had been riding with Mazeppa," a reference to Lord Byron's 1819 poem *Mazeppa*, in which the title character is punished by being tied to a wild horse that is set loose to run freely. The writer then contrasted that image of a long and speedy journey, fraught with potential danger, with the appearance of the "quiet, grimy engineer, turning already to his tobacco and his newspaper, and unconscious, while he reads of the charge at Balaklava, that his life is Balaklava every day."[119] Here, the author made the reference to the Crimean War's Battle of Balaclava—fought just five years before this guide was written and immortalized in poetry by Alfred, Lord Tennyson—to underline how travelers had made routine what in fact could be a dangerous undertaking. At a moment's notice, a journey could turn deadly. Such frightening reality, juxtaposed with the engineer's nonchalance, suggests that Americans lived with and accepted the risk of travel.

For railroads, safety in this era was largely dependent on time management. Even with the invention of the telegraph in 1844, telegraphed orders were not a central part of safety. In part this is because using time management alone was successful in preventing accidents for the first few decades of railroad operation. Historian Ian Bartky notes that of over 175 nineteenth-century train wrecks, "faulty timekeeping caused only three."[120] Railroads were so successful in managing time that corporations were indeed skeptical that the telegraph would improve their operations. These early timetables were a "safety guide for each engineer and conductor." There were no nationally standardized time or time zones in this era, and even nearby cities could have slightly different times. Nevertheless, railroads managed to cooperate well enough to keep things operating safely. Major, deadly accidents did not increase until the 1850s, when the railroad network had become larger and denser, so time management alone could no longer suffice.[121]

Despite an overall acceptance of the dangers, Americans were not cavalier with their lives. Plenty of Americans were concerned about the safety of steam transit. In 1838, Angelina Grimké wrote to Theodore Dwight Weld, imploring him to be careful: "Take thy seat in a car *far off* from the engine."[122] Sometimes, the surrounding view had the potential to heighten the sense of danger. Looking back at some mountains once he had safely crossed them by train, Charles Titus was amazed that he "felt so easy & safe in the passage. At some places it was fearful to look down the deep & almost perpendicular precipices on one side, & up to the ragged Bluffs towering almost to the sky on the other."[123]

In 1851, Lincoln Clark revealed his concerns about railroad travel to a relative: "We were much relieved and gratified to know that you had accomplished your journey without casualty or suffering. There are so many accidents happening by railroad that we never feel quite at ease when our friends are on that lightning track until we know that they are safe at the end."[124] When accidents did happen, they could be the talk of the town. In 1841, John Williams Gunnison in Green Bay, Wisconsin, wrote that "the talk is solely on the loss of the [steamboat] Erie which was burnt on the Erie Lake last week & with a hundred & sixty persons, destroyed." Gunnison noted that "public confidence will be shaken" in this "convenient & powerful" mode of travel "unless more care is taken in its management."[125] Some advertisers exploited these fears. In an 1860 advertisement for life insurance, the ominously titled section "Testimony from the Dead" included the story of a minister "who was killed on the railroad at Gasconade Bridge" but whose family received $6,443.55 in compensation thanks to the insurance plan.[126] The message was clear: disaster could strike anyone at any moment. Even a pious man of God had to be prepared for the worst.

Weather could contribute to dangerous conditions. In 1840, Robert Habersham described an experience where the snow was so deep that the train could not continue. Some people on the train resolved to get out and walk along the track. The train eventually broke free and apparently ran over one of the men walking on the track. "By a merciful Providence," Habersham reported, the victim was only "somewhat bruised about the face."[127] In 1856, a snowstorm in Illinois buried the tracks of the railroad in that state "twenty feet in places," according to an engineer. After a week of shoveling, the trains were dug out, only to be buried again by another blizzard.[128] As these examples have highlighted, weather had the potential to increase risk for both passengers and company alike.

Steamboats presented their own unique danger if two steamboat captains decided to race each other. Steamboats "excited the imagination because they were so fast."[129] Antebellum narrators recounted how the temptation of a race could lead to reckless behavior from captains. The goal of beating a rival to bring honor and glory to their ship could lead them to disregard the safety of their passengers. Ann Archbold described such an occurrence in 1848: "One pleasant evening we were all sitting at our ease, conversing about matters and things in general, entirely unconscious of danger, when all of a sudden the boat struck a dreadful sand-bar with such force, that every stick of timber within her groaned, and the men were thrown from their chairs on their faces, the shock being most severely felt in the gentlemen's cabin." Panic descended onto every part of the boat, leading to a disorienting sensory experience. In addition to the physical jarring of the boat, the sounds surrounding Archbold changed: "Screams of terror and dismay were ascending from many, in the lower deck."

Archbold herself credited God with keeping her "mind in a state of perfect composure," confident that she would "rise to life and immortality, when the struggle of life was over." After the panic had passed and the boat did not sink, Archbold spoke to the captain. The captain admitted that "he knew that the boat was going to strike, before she [the boat] did, but was sailing so rapidly, that he could not save her from this." And why was he going so "rapidly"? Because he was racing another boat. His boat won the race, since the competing boat had struck a third boat and had to stop. The captain boasted that "the race was worth at least a thousand dollars to him, as it would raise the character of his boat."[130] With such an incentive structure in place, passengers had every reason to worry that captains would be inclined to take the risk. In 1852, *Godey's Lady's Book* spoke out against the need for steamboats to best each other in terms of speed when it commented on the death of someone on board the steamboat *Henry Clay*: "None has been more lamented among the multitude there sacrificed to the reckless spirit of emulation which gives the swiftest steamboat its popularity, and therefore its conductors their power over human destiny."[131] Surely there were dangers enough without adding to them through unnecessary racing.

Much like other facets of their travel experiences, travelers related accidents in their letters and journals. In 1838, Andrew Leary O'Brien recounted this experience out of New York City. A conductor told passengers to "look out," intending that to be a warning. But one passenger interpreted that as an instruction to stick his head out the window. When he did so, "a bridge post took hold of him across the shoulders, & dragged him out." O'Brien "caught him by the coat collar & held on with all my strength, till he passed the post." The man appeared to be quite hurt, but the "next day we discovered no bones were broken."[132] In 1849, Lincoln Clark described a railroad accident:

> The whole bulk of passengers in each car was thrown like a wave of the sea and dashed upon the floor; the seats themselves snapping from their fastenings like pipe-stems followed their occupants, and all was a scene of confusion:—

some came out with gashes in their heads; some with their mouths and lips cut, some with their teeth almost knocked out; the blood dripping freely upon the snow:—none I believe received lasting injuries unless it was the man in charge of the mail and a little child. I escaped without the slightest bruise or harm—for which I believe I did humbly thank God most merciful.[133]

In his account, Clark likened the accident to a natural event ("like a wave of the sea") and thanked God that injuries were not worse. Caroline Barrett White also considered God when she wrote of a "melancholy accident" at the celebration for a railroad opening in which "a young man about twenty years of age was so badly injured as to cause his death in two hours, several others slightly wounded. Surely 'in the midst of life we are in death.'"[134] White was reminded that God could call people "home" at any moment.

Of course, not all the dangers of travel came from technology or the reckless actions of the crew. There was also the danger from fellow passengers who might attempt to take advantage of inexperienced travelers. In 1857, an illustrated guide alerted travelers to the dangers of New York City. The book warned against pickpockets, saying that they "are never at home except in a crowd; then only can they work to full advantage and without exciting suspicion; and therefore whenever there is a rush of people to any particular spot, then are these gentry the first on hand. They are always in full feather at Steamboat Excursions, Grand Railroad Jubilees, Conventions, State Fairs and public shows of all kinds." Women pickpockets, in particular, "frequent the cars, omnibuses, and stores."[135] Women were particularly dangerous in this respect because they were probably least suspected of being potential criminals. A particularly cunning method of robbery was to soak a handkerchief in chloroform, wave it in front of the face of the victim, and rob the person after they passed out. The book noted that this was done "in railroad cars, upon both lady and gentleman passengers, by both male and female scoundrels; indeed, upon a night train the thing would be most convenient."

Trains in general were seen as good places for thievery because of the constant physical motion of the train. A pickpocket could naturally bump up against someone due to "the passing, re-passing, and jostling" of passengers on the train. To avoid pickpockets, readers were advised *"to keep out of crowds."* Thieves would "sometimes get up sham fights at the steamboat landings and railroad depots, and to give the thing a real look, stand regular licks and bloody noses; in the rush and pressure to see the fun, the accomplices of the combatants usually contrive to do a very handsome business."[136] Savvy and experienced travelers understood the multitude of risks, even the seemingly innocuous ones.

Travel guides also advised readers about other types of dangers, like "Bogus Ticket offices," which, although "few in number," would offer "the sale of bona-fide tickets at from 25 to 100 per cent. advance upon the regular rates; the palming off of second for first-class tickets, at the full price of the latter; and the taking the full price of a through passage, and giving a ticket for but part of the distance." The chief goal of this trickster was to ensure that the "plucked individual gets his feet clear of New York mud, and he once more breathes the pure air of heaven, his reason returns, and then has too much sense, or too much shame, to return and seek what he has so unwisely lost."[137] Distance between the mark and the criminal mattered, but so too did the shame of the person taken in by the ruse.

As a sign of how accepted the dangers of steam travel were in American life, accidents found their way into the arts and other forms of cultural expression. Poetry could express shock and sadness at the loss of lives. A poem written after an explosion in 1828 "on the death of 30, or 40 men Killed in a Steam-Boat on the Mississippi River, by the bursting of the Boiler," included the gruesome stanza "One thundering crash now Rent the air!/And one fierce cry of Keen despair,/And mingled Horrors blending loud,/and groans & shrieks Ran thro' the crowd./Th' imprison'd fires had swept the decks,/& strow'd the Scene with Human wrecks/O happier were their milder doom,/Hurl'd Sudden to a watery tomb."[138] A poem in 1859 bemoaned the impunity with which accidents were going unpunished: "Sometimes two trains will run afoul, then the mischief is to pay;/Nobody knows the cause, I'm sure—the guilty, none can say./They gather up the wreck of life, inquire the wounded's name;/An inquest's held, the verdict is—That 'nobody's to blame!'"[139]

In addition to poetry, disasters were also recorded in artwork. The steamboat *Lexington* suffered an accident in 1840. Lithographer Nathaniel Currier happened to be in a newspaper office when he learned of the disaster, and immediately "had the brainstorm of illustrating the extra that editor Benjamin Day planned to issue." He set his employees to work, "supplying them with fresh details as they became available from eyewitnesses. Within three days, the illustrated news extra hit the streets. It was an unprecedented feat." The artwork, published again separately from the newspaper, ended up becoming popular nationally, and "Presses ran day and night to keep up with demand."[140] Images could demonstrate the danger of accidents; Currier's representation of the *Lexington* featured people in the water grasping onto objects while the ship itself was consumed in flame.[141] Similarly, the 1855 accident on the Camden and Amboy Railroad was reproduced by an artist. The image shows several cars destroyed and the passengers strewn alongside the destroyed train.[142]

How, then, did Americans inure themselves against the dangers of steam travel—how did they manage knowing that it was Balaclava every day? In part, as we have seen, this is because most railroad accidents were not major accidents with multiple deaths. Therefore, antebellum passengers were capable of separating the dangerous accidents that led to loss of life from those that were mere inconveniences. In 1851, Lurana Y. White Chase wrote in her diary of an accident that happened but did not delay their journey too much: "The cars run off the track, which detained us a short time, no one was injured, we were soon puffing along at an immoderate rate."[143] Chase was bothered by the accident, but it did not sour her on travel via steam. Additionally, there was a widespread trust that—as with all things—God's will was at work and all lives were in God's hands.

Americans accepted accidents as the cost of receiving the benefits of steam transit. A railroad guide noted that Americans considered the accidents but did not shy away from travel: "We have seen a whole car full sit quietly and read an account of a terrible accident that sent half a hundred poor souls into eternity, without once reflecting that they were liable to the same fate, the next moment." It seemed incredible to comprehend, but the guide continued: "Why should one torment himself before his time? If the purest water we drink was examined through a microscope, half of us would die of thirst before we could swallow the animalculæ that are swimming there."[144] Risk then, was all around. But Americans wanted the benefits of speedy steam travel. As the *National Era* wrote in 1858, "The travelling public love to ride in one of these 'Lightning Runs,' which whizzes them over the track at a rate as dangerous as it is delightful. They love a speed which grinds the rails, racks the axles, strains the timbers, renders it impossible to stop in time to avoid accident, and insures death and destruction when accident does occur. So the danger is incurred to please the passengers."[145] A commentary on the *Lexington* steamboat disaster noted the trade-off that people seemed to be making: "A diminished rate of fares and a superior degree of swiftness seem to have the power to stifle the sense of danger."[146]

Steam transit was cheap, fast, and exhilarating. Most people never experienced a deadly accident. The odds seemed to be in the rider's favor. Despite accidents and the inherent danger, the number of passengers carried on railroads grew steadily throughout the antebellum era. In 1839, railroads in the

United States served ninety million passenger miles, and this grew to 1.8 billion passenger miles in 1859.[147] Steam transit had its risks, but within the developing culture of American travel those were risks that Americans were willing to take. The deaths by steam transit challenged the Good Death by rendering death a possibility at any moment. The violence of the Civil War, however, would soon present a far greater challenge as Americans died as never before. Steam transit accidents had demonstrated the possibility of being robbed of the Good Death, and the Civil War would push that concept firmly into the mainstream.[148]

The willingness to accept risk was simply one cultural evolution in how Americans traveled in the antebellum era. At the beginning of the nineteenth century, Americans worried about where they might spend the night if the weather turned bad and they had not yet ridden a horse to their destination. By the eve of the Civil War, Americans were hurtling through space at previously unimaginable speeds. In the intervening decades, travel became commodified. No longer did trips have to be designed and implemented by each individual traveler. One could simply purchase a ticket from the ticket counter, and the corporation would take care of the rest. Over time, passengers created their own culture of how to act on railroads and steamboats in order to manage the process of spending a large amount of time with strangers. Some of this culture could be learned in books; other aspects were enforced in the moment by other passengers. Travel by steam excited all of the senses, from the smell of the rising smoke to the jostle of a bumpy train ride or the shudder of a steamboat hitting a sandbar. Over the antebellum period, Americans familiarized themselves with travel by steam and its trade-offs as these modes spread across the country, in so doing creating a culture of travel. All of the facets—commodification, sensory experience, acceptance of risk—helped naturalize travel for Americans, so that in general they did not question whether travel was too dangerous.

CHAPTER THREE

The Arts

Steam transit created a new physical presence on the American landscape, but the influence of steam quickly expanded into the cultural realm. Steam transit occupied mental space for Americans as well as physical space; in fact, as steam became more prevalent and popular, its cultural impact stretched beyond its physical reach. Actual railroads and steamboats took time to expand throughout the United States, but the cultural reach of steam transit required only the availability of written descriptions or visual depictions, which took many different forms. Railroads and steamboats became ready metaphors for writers, in both published writing and private correspondence. Humorists took aim at steam transit as a subject of mirth, and a dark humor developed around the threat of accidents. Music composers took steam transit as their subject, writing music that honored the growth of rail lines or steamboats, even trying at times to imitate the sound of steam transit in their compositions. Images of steam transit were omnipresent on something as common as currency and "higher" expressions of art as well. Finally, steam transit generated its own unique literature—guidebooks and timetables that assisted travelers or provided a written picture of the route for travelers and those who stayed at home.

When looking at all of this cultural production, only the last category of guidebooks is directly tied to the travel experience. The remaining groups demonstrate how thoroughly steam transit became embedded in American culture. People did not have to ride in a train or steamboat in order to observe or participate in the larger cultural conversation. Even people who lived far from steam transit could read about it, laugh at the jokes or cartoons, or play the music on their piano, all of which made steam transit part of their world despite their never having boarded a train or steamboat. This rich cultural output demonstrates how Americans familiarized themselves with stream transit perhaps before they had the opportunity to experience it in person. These cultural expressions naturalized steam transit, making it easier for people to accept and understand steam transit when it finally arrived in their town or when individuals first interacted with steam transit in person.

Metaphors

Metaphors relating to steam transit abounded in the antebellum era, emphasizing a variety of aspects of steam travel, and were found in both public press and people's private writing. The fact that metaphors were used by all demonstrates how deeply they were held in the American mind. As we will see, speed is unsurprisingly a common metaphor, but writers also make reference to pressure—a significant metaphor that resonated with Americans in this transformative and changing age. Even if the public did not grasp the finer points of physics, the constant references to pressure illustrate that the public knew at least one of the basic facts of steam operation.[1]

One example of a pressure metaphor comes in Thomas Shreve's 1835 story "The Young Lawyer." In the story, one character offered the lawyer an alcoholic drink in order to rouse his spirits. He justified the offer by saying, "A man's like a steamboat—he may have good works aboard, but damn the bit he goes ahead, till he gets steam up." The lawyer declined the drink by extending the metaphor: "Thank you, Joe; I've got as much steam about my upper works as I can navigate well under. If I should take any more aboard, some of the flues might collapse, and the boiler burst, or some other mishap come over me."[2] Steam was the perfect way for the author to illustrate both the offer of the drink and the polite refusal. The lawyer needed the energy, but too much pressure would be dangerous and cause physical harm. A burst boiler would be familiar enough to readers for them to understand that the lawyer wanted to avert disaster. In 1848, the *National Era* used another metaphor about a boiler, speaking about people who draw attention to themselves: "Some men are constructed on the high pressure principle. They puff and blow, and heave and snort, like one of our Western steamboats. They 'go ahead,' it is true, but they take care to make a great ado about it, and cannot bear to move without advertising the world of the fact."[3] Any person who had heard steamboats would appreciate the similarity to a pompous blowhard who drew attention to himself.

In the *National Era*, a writer, using a steam metaphor the following year, complained about the large amount of "bad novels" read by young women. He charged that people who read such fiction never gained any education or deeper thought thereby. Rather, these were works that put the reader "under the steam pressure of sentimentalism." Under this pressure, readers "took a literary railroad ride every day, made up of dash, rush, a little zest of alarm, whirling landscapes, a collection of strange faces, an occasional shriek, or a loud laugh, giddiness, fatigue, and finally the whole is dumped down at the regular depot, just as everybody expected when they took their tickets for the trip." The ride was long and perhaps exciting, but there was no change in the reader at the end: they gained nothing from the experience.[4]

The most common way in which steam transit was used as a symbol in the antebellum era was unquestionably for speed, or the pace of American life. The *Crockett Awl-Man-Axe* mused in 1839, "'We are born in a hurry,' says an American writer, 'we are educated at speed. . . . Our body is locomotive, travelling at ten leagues an hour; our spirit a high pressure engine.'"[5] *Railroad speed* was a widely used phrase for moving quickly, which highlights the fundamental changes to travel that steam delivered. It appeared in commentary on a marriage proposal from 1853: "M. J. Hedges, a conductor on the New York and Erie Rail road, proposed to Miss Henrietta Converse, at Elmira, at 4 P. M., was accepted immediately and at 7 the same evening the happy and expeditious pair were married by the Rev. Dr. Murdock. That might be called courting and marrying at railroad speed. Fast country, this."[6] A guidebook for the Eastern Railroad warned that at the stop at North Berwick, Maine, there was only five minutes to get food, so it "must be swallowed with rail-road speed."[7] A book reviewer in 1858 noted that the book under review "seems to have been written at railroad speed, . . . so rapid are the changes in scenery and incidents."[8] Throughout the antebellum era, on a range of topics, the railroad was a metaphor for fast living.

Other parts of the steam travel experience were invoked by writers. Women's rights advocate Elizabeth Cady Stanton relied on another part of the travel experience—waiting at the depot—to characterize the plight of women in 1859. She wrote, "The woman who has no fixed purpose in her life is like a traveller at the depot, waiting hour after hour for the cars to come in—listless, uneasy, expectant—with this difference, the traveller has a definite object to look for, whereas the woman is simply waiting for something to 'turn up.'"[9] Stanton urged women not to be passive, and the metaphor of listlessly waiting for a train plainly illustrated how not to act. Writers also used steam transit to show that time had passed within the story or to mark the time in the story as distinctly different from the present day. Comparing and contrasting the time before and after steam transit existed made it easy to show the reader that a significant barrier had been crossed. When writing of nostalgia for "the days of harmonious village and happy farmstead," Harriet Beecher Stowe once referred to that period as "the ante-railroad times."[10] Metaphors could also address issues such as demeanor, as the following examples illustrate. An 1838 etiquette guide used a steamboat to make a point about shaking hands: one should "press his hand with gentle warmth, not pulling it as if it were the rope of a steamboat-bell."[11] Another etiquette author wrote, "Habits may be likened to a rail-road, on which the carriages move smoothly and easily." Poor habits would make it easier to do wrong things, just as a poorly constructed road would throw the carriages from their tracks.[12]

Railroads were relentless as they moved along the track, which was also noted in literature. In Herman Melville's *Moby Dick*, Captain Ahab declares, "Swerve me? The path to my fixed purpose is laid with iron rails, whereon my soul is grooved to run. Over unsounded gorges, through the rifled hearts of mountains, under torrents' beds, unerringly I rush! Naught's an obstacle, naught's an angle to the iron way!"[13] Ahab was a ship's captain on the wide and boundless ocean, a mode of transit that allowed travelers to go in any direction. Nevertheless, Melville put into Ahab's mouth the metaphor of the unerring and unwavering train in order to underscore the character's consistency of purpose. In its unshakeable path, the mode of transit Ahab used as a metaphor was the opposite of the one he commanded but spoke very well to the character's determination. In her novel *Uncle Tom's Cabin*, Harriet Beecher Stowe described one of her characters as being "as inevitable as a clock, and as inexorable as a railroad engine."[14]

Other writers latched on to railroads as a metaphor. In 1852, a periodical contained an item called "The Railroad to Ruin." It was a pro-temperance piece and indicted alcohol consumption with an extended metaphor: "Surveyed by avarice, chartered by county courts, freighted with drunkards, with grog-shops for depots, rumsellers for engineers, bar-tenders for conductors, and landlords for stockholders. Fired up with alcohol and boiling with delirium tremens. The groans of the dying are the thunders of the trains, and the shrieks of the women and children are the whistles of the engines. By the help of God we will reverse the steam, put out the fire, annul the charter and save the freight."[15] This metaphor linked alcoholism to multiple aspects of the railroad—from survey to operation—in an effort to portray it as comprehensively as possible. The result painted a stark picture for the reader that showed just how intimately all of these parts were connected into a disease, illustrated as a machine that entrapped drunkards.

While the range of written metaphors for steam transit in the public realm was impressive, far more indicative of the pervasive impact of steam transit in the American mind is the fact that so many writers used the imagery in private correspondence and diaries. Such metaphors were not just the provenance of those writing for the public. People from all walks of life referred imaginatively to steam transit in their private correspondence as well. Dolley Madison wrote to a friend in 1833, "How delighted I should be, with a Rail Road, which would convey me, in a few hours to you and many estimable friends near you."[16] Even at that early date, the railroad held an appeal for speedy travel and could help friends stay connected. In 1839, Caroline Healey Dall wrote that her journal was her "safety valve—and it is well, that I can thus rid myself of my superfluous steam."[17] George Templeton Strong wrote in his diary that he was "getting to be a kind of professional steam engine" because he was working so hard.[18] Frederick William Seward recorded that he heard the following about John Quincy Adams in 1843: "A characteristic expression of a steamboat captain with whom he travelled illustrated the popular feeling. He said: 'Oh, if you could only take the engine out of the old Adams, and put it into a new hull!'"[19] It was high praise for Adams's spirit and expressed the desire to see him keep working.

In 1843, Dexter Russel Wright compared himself to a steamboat in an extended metaphor written in his diary:

> My course is like that of a steamer. She glides noiselessly along without a sign of exertion, as though it were hostile to her nature to stand still. But if we look a little farther—if we pass within her beautifully adorned exterior and inquire what makes it "plow the waters like a thing of life," we discover what incessant and giant efforts her engine is putting forth to urge her onward. Thus while all is gentle, easy, and placid without, there is a mighty struggle within. So, with me; if I have ever accomplished anything in life, it is by much boil and perseverance. Though I may seem to move along the stream of time smoothly, yet my mind is constantly exerting all its energies to overcome difficulties and surmount obstacles.[20]

Demonstrating his familiarity with how steamboats functioned, Wright then chose the steamboat as the perfect metaphor for his life: calm on the surface while churning turbulently below.

The equation of railroads and speed also entered private conversation. One letter writer sent a second letter before awaiting a reply and justified his actions by saying, "In these Rail Road times I go ahead and that is all the apology I have to make."[21] In 1853, Henry Blackwell wrote to his brother about a woman he hoped to see again. But he was concerned that this would not be possible, "as she is travelling around—having been born locomotive I believe."[22] And in a letter by Anna Rebecca Young, she chastised her correspondent's poor handwriting while still being glad to receive the letter: "it was most welcome in despite of words running into each other, like railroad cars have of late been doing."[23] Such examples of steam transit in private correspondence show that Americans saw steam transit as more than an economic proposition but used references to transit to communicate feelings and opinions to their friends and family. It was a familiar set of references that would instantly communicate meaning to the reader.

Steam's impact was not simply in bales of cotton carried or miles traveled. The infiltration of steam into the language of ordinary people and correspondence demonstrates the impact of steam transit on the American imagination. Average Americans may not have fully comprehended the physics involved in steam transit, but they understood enough to make use of the metaphor of pressure. They could tell that steam transit was a meaningful metaphor for speed. This alteration of language was not just the province of railroad promoters, newspaper editors, and civic leaders but can be found in the written diaries and letters of Americans throughout the antebellum era. In this way, the influence of steam stretched into thought and language of many Americans, even those far from the tracks and wharves. The concept of steam permeated the culture to express anew and in a more vigorous fashion concepts and emotions familiar to all.

Humor

Steam transit became a quick target for humorists in the early nineteenth century. Cartoons, jokes, puns, and comic stories on the topic of steam abounded in this time period. All of these objects illustrate, again, how steam transit was accepted and interwoven in American life: even if people had not yet boarded a train or steamboat, they could read the humor and laugh along at the absurdity of a comic situation. The possibilities for such situations were many: the behavior of people on a train or steamboat, the overzealous claims of railroad promoters, and the amazement that the speed of transit brought. The more macabre jokes underscored with a sly wink how dangerous transit could be. This gallows humor constituted a coping mechanism for Americans as they confronted both their desire for travel and the reality that a deadly accident could strike at any moment.

One particular form of antebellum humor was the range of comic almanacs that poured forth from printers' presses. These books included a variety of jokes and cartoons, puns and anecdotes, and many of them featured steam transit. As historian Constance Rourke wrote in her study of humor in the United States, "These fascinating small handbooks yield many brief stories and bits of character drawing not to be found elsewhere; more than any single source they prove the wide diffusion of a native comic lore." The almanacs were published in Eastern cities but also found publishers farther west, centered around characters such as Davy Crockett.[24] In all, they offer a range of humorous anecdotes about steam.

Some early jokes were completely fanciful, playing on the astonishment that steam transit brought when it was a novelty. A book of humorous anecdotes published in 1839 included several demonstrating this amazement. One anecdote reported that if a person put a vial of the "concentrated essence of the sublimated spirit of steam" in their pocket, then it would be possible to travel "at the rate of fifty miles an hour." And if a person ingested "three drops" of the same concoction at bedtime, then "in the morning you will wake up in any part of the world you like." At the time that this story was written, the concept of fifty miles an hour would have seemed like a laughable exaggeration. But the underlying message was that steam brought with it incredible speed with limitless possibilities. Another anecdote reported that a steam engine had been built that moved "with such expedition, that the eye cannot catch them in their transit. The passengers enter the carriages, and are set down at their journey's end, before the moment in which they entered them is concluded. They find themselves in the place in which they wished to be, without being at all able to conjecture how they could arrive at it." Again, exaggeration here served a purpose, hinting at possibilities of travel. People could laugh at the exaggeration, all the while appreciating the changes that steam had wrought. A third story in the collection implied that steam's power had moved to other contexts. The anecdote claimed that when asked about typographical errors, a newspaper editor had offered as an excuse the fact that since individual pieces of type had "been so often used in notices of railroads and steam-boats, they have *the principle of locomotion* so thoroughly infused into them that they are continually jumping up and down, and not unfrequently alighting in places appointed for others."[25] Steam transit's power here extended to the very printing press itself, asserting its influence on the letters.

Other stories in the antebellum era also played on the fanciful transposition of steam travel to other contexts. A humorous story published in the *Boston Pearl* in 1835 featured a man attempting to escape from advocates of phrenology, the belief that the size of the cranium predicted mental acuity. Adherents of this pseudo-science walked up to him to examine his "remarkable head." He decided to escape them by traveling to Little Rock, Arkansas. He hoped to live in peace there, but the story concluded with him writing that he had seen a boy posting signs that lectures on phrenology would be delivered in the town. "Verily the science should be applied to the driving of locomotives on railroads," the author concluded, "for it out-travels steam."[26] The popularity of this "science" led it to travel rapidly throughout the land; the only sensible point of comparison for its speed was steam. Another humorous story reflected on the destruction of livestock by trains. The author joked that a patent had been put in for "a locomotive pickling apparatus for salting down the cattle killed on railroads. The machine will prepare the beef as fast as it is delivered in the rough by the cow-catcher."[27] The destructive power of the train was comically converted into a welcome efficiency. Yet another story using exaggeration was written about a "WEATHER INSURANCE OFFICE." In this story, the author discussed how different people could ensure that they got the weather that they needed. One such group of customers were "Captains of Steam-Boats," who "may be accommodated for an 'Excursion' with a remarkable fine day, at a moderate premium."[28] All three stories were based on whimsy or fantastical exaggerations, but all had their root in the impressive power of steam transit.

Steam, rail, berth: many words associated with steam transit lent themselves to puns, and groan-inducing wordplay featured throughout the antebellum era. Oliver Wendell Holmes protested against puns in general in 1857, using a railroad metaphor to do so: "People who make puns are like wanton boys that put coppers on the railroad tracks. They amuse themselves and other children, but their little trick may upset a freight train of conversation for the sake of a battered witticism."[29] If the comic almanacs and other humorous writing are any indication, writers upset freight trains of conversation with abandon during the antebellum era.

In 1843, the *Old American Comic Almanac* included a tale in which one Irish immigrant to the United States asked another if he had ever seen a railroad. The latter responded that he had never seen one, but had heard one, since his wife had been *"railin' away"* since they wed.[30] In this instance, the immigrant, having never before seen a train, used this metaphor to denigrate his wife. Another pun likewise relied on ethnic stereotyping, when an Irishman dragged a plank toward a steamboat. The boat's captain asked what he was doing with the plank. The Irishman retorted indignantly that the captain had himself said "Get a board," and thus he was merely following instructions.[31]

One comic book from 1853 included the following two steamboat puns. In the first, a slow steamboat was called the Regulator, because "all other boats go *by it*." Playing on the meaning of "go by"—both to pass and to use as a point of reference—the pun mocked the slow speed of the boat. In the other pun, an "old lady" read a newspaper article that a steamboat "had twelve berths in her ladies' cabin" and then declared, "What a squalling there must have been!"[32] Here playing on *berth/birth*, the pun poked fun at the old lady's mistake. *Fisher's Comic Almanac* demonstrated that even youth could get into the act: "'What are the ways of Providence?' said a Sunday School Teacher in this city to a little boy in his class the other day. 'Railroad to Boston and Steamboat to New York,' answered the urchin. This was mixing Theology and Geography with a vengeance."[33] Through humor, steam transit's reach even included Sunday school. Sometimes the puns were visual. In 1839, *Turner's Comic Almanack* published a cartoon of a man walking and sweating profusely, with the caption "Walking by Steam, to get a glimpse of the 'Great Western.'"[34] And in 1852, the *Carpet-Bag* printed a cartoon of a few men sleeping in a passenger car, with the legend "Spruce Railroad Sleepers. Railroad contractors supplied with any desired quantity, to be delivered at the terminus."[35] Here, the sleepers were not crossties to support the tracks but passengers dozing on a trip—and these were spruce, dandy fellows, not spruce wood. These images took common formulations and used them to laugh at their subjects (fig. 3.1).

Very quickly, antebellum humorists picked up on the macabre. The dangers presented by steam travel were real. Humorists used this potential danger as a punchline, and they also employed their humor against the companies. In 1835, *Elton's Comic All-My-Nack* included a picture titled "The Bowery Locomotive, or the Pleasures of a Rail-Road." It showed an engineer reading a newspaper as his train ran over several people.[36] The engineer's clear indifference led directly to the deaths of bystanders, and the picture indicted the railroad for its apathy for the damage it caused. Another comic almanac included an image of a steamship exploding and all the passengers being tossed in the air; the steamship bore the name "High Flyer." The drawing

SPRUCE RAILROAD SLEEPERS.
Railroad contractors supplied with any desired quantity, to be delivered at the terminus.

Figure 3.1. A pun on *sleepers* links railroad construction to those who snoozed during their journey. *Carpet-Bag* (October 16, 1852). Catalog record 17847. *Courtesy, American Antiquarian Society.*

played on the aspirations of steam transit—to "fly" ahead with all possible speed—with the dangerous possibility of explosions. Below the drawing was written this text: "The aid of *steam* applied to boats and locomotive engines on rail roads goes a great way towards 'annihilating space' as well as *annihilating* those who travel in these conveyances. The good old sociable manner of travelling is fast going by; and travellers had sooner run the chance of being blown 'sky-high' than be one hour longer in travelling five hundred miles."[37] The story and image thus tapped into a reader's concerns that traveling quickly was not an unalloyed benefit.

Another dark joke about danger on a steamboat appeared in 1839 when a nervous passenger paid his fare of one dollar and asked "if there was no danger of being blown up." The response was clear: "No sir, not in the least; we can't afford to blow up people at a dollar a-head."[38] The profit motive was drolly credited here for preventing accidents (due to the costs involved) rather than causing them (by cutting corners). That same year, the *Old American Comic Almanac* featured a train on its back cover. The train was labeled with the title of the almanac and was being driven by a skeleton laughing and pulling at the bell. The engine pulled two cars full of people with smiling faces. The train was crushing apparent rivals of this publication: "Comic Manual," "Comic Sheet," "Jokes," and others. Above the image the text imagines a conversation between a passenger and conductor. The passenger asks, "Pray, Mister do you call this a first rate carriage because it goes double-fast?" The chilling response: "No marm, it's because we puts it behind to be blow'd up last."[39] Other comic almanacs also played on the danger of a journey and fears of travelers. In 1858, a joke noted that one man was not worried about steam explosions since "he was so used to being blown up by his wife, that mere steam had no effect upon him."[40] This joke made an equal target of the wife and steam transit, with both threatening to "blow up" the male victim.

Young America's Comic Almanac for 1857 took a few jabs at railroads, both poking fun at the passengers and pointedly noting how companies evaded responsibility. In one cartoon, a skeleton wears a conductor's hat, with the caption "The Conductor of the Train" (fig. 3.2). In the accompanying story, an "elderly gentleman" on a train bids a melodramatic goodbye to his wife and six children. The wife cries out that her husband is "going to leave us forever," and the man despairs that he is going "twelve miles in the country, and I know I shall never return." When the train pulls out, the family members on the platform either faint or "go into hysterics." This family was plainly overreacting to the danger of the journey. But the almanac also turned its attention to the companies themselves. In a series of "Did you ever" statements, there is this listed: "Did you ever know a jury to find anybody guilty in case of a railroad smash, or a steamboat accident?" This joke expresses frustration that transportation companies had a degree of impunity and were not held liable for their accidents.[41]

Beyond the fanciful yarns about the powers of steam, wordplay around transportation, and gallows humor about the dangers of transit, there were plenty of jokes and humorous tales that took railroads and steamboats as their setting. As historian James Justus has noted, for humorists of the Old Southwest, the steamboat in particular was an appealing stage on which to set humorous stories. "The social hall and boat decks were *stages*," Justus argues, "both for the enactment of little dramas and their reenactment in taletelling. The steamboat was what the humorists most cherished—a theater of possibilities."[42] Steam transit brought strangers together briefly for shared experiences, and they proved to be the perfect setting for stories.

Several stories poked fun at steamboat captains. In 1840, for example, the *Colored American* included a story about a "Yankee pedlar" who was trying to

The Conductor of the Train.

Figure 3.2. A morbid reminder of the possibility of death on rail transit. *Young America's Comic Almanac for 1857* ([New York]: T. W. Strong, [1856]). *The Library Company of Philadelphia.*

travel from New York to Poughkeepsie without paying. Just before the steamboat arrived at Poughkeepsie, he went to the captain and apologized that he had no money but wished to travel all the way to Albany, farther along the route. The "indignant" captain then put the peddler out of the boat at Poughkeepsie, just as the peddler intended.[43] Another captain was at the heart of a story in the *National Era* in 1847. While traveling on a steamboat, he was conversing with passengers, but it was quickly evident that he was only "'in town' as long as the conversation was about steamboats." When a woman passenger asked him what he thought of the "immortal Shakespeare," he immediately responded that "she burns too much wood, draws too much water, and carries too little freight."[44] The captain was so taken with steamboats that he could not talk of anything else, and assumed the question was about a steamboat. Another story told of a steamboat, traveling in a dense fog, whose captain repeatedly and unknowingly bought wood from the same woodyard. He complained about the price each time. The last time, the owner of the yard agreed to lower the price slightly, *"as you're a good customer!"* which finally revealed to the captain the error he had made.[45]

Passengers were also the target of humorous stories. Sometimes these stories highlighted the fluidity of identity that steam transit made possible. In a story from 1859, a young couple wanted to get married, but their parents would not permit it. So they boarded a steamboat in Kentucky and asked around for a clergyman. Some fellow passengers—"lovers of fun"—found someone to serve as an "impromptu parson," and the counterfeit priest performed the ceremony. When the couple was about to retire to the stateroom, the captain, "thinking the joke had proceeded far enough, intervened and revealed the imposition, in time to prevent any actual damages." The enraged groom-to-be made a "fruitless search for the fictitious clergyman," who had disappeared. The couple were then wed the next day by a legitimate parson.[46] The couple had seen the steamboat as an opportunity to escape the restrictions that they experienced on land, and they in turn were tricked by the other passengers.

For some authors, the wonders of steam technology offered the opportunity to mock a "rustic" character. In 1834, *New England Magazine* included the tale that "Jonathan Jolter" left Vermont to see the wider world. His assessment of a train upon first sight: "This must be the sea-snake, from an overland tour/Returning, pell mell, his own ocean to find,/and dragging a horrible earthquake behind." Despite being disturbed at the sight, Jonathan screwed up his courage and declared, "No Yankee knows fear, that e'er grew on the soil." But a further trip to Boston and more encounters with steam convinced him that "the ubiquitous steam 'must rise from the bottomless pit itself.'" For this Vermonter, steam could only come from hell. According to Cameron Nickels, focusing on such characters "calmed fears about cultural change and modernization because they judged rustic values and experience as hopelessly, because laughably, out-of-date."[47] By mocking the rube, readers stabilized their own feelings and concern about technological change.

A similar story was published in 1859, titled "Greenhorn's First Ride." The greenhorn boarded a train, but after the ride started he decided that he wanted to get off. At that moment the train was crossing a gorge on a bridge. The young man dashed out onto the car's platform, and "seeing the earth and treetops beneath him, he fainted and fell. Directly he came to, and looking up at the conductor, who stood by him, he exclaimed with a deep sigh:—'Oh lordy, stranger, *has the thing lit?*'"[48] The greenhorn thought that the train had taken flight, exposing him to ridicule from the reader.

By the last decade of the antebellum era, railroads were unquestionably symbols of progress, and stories could poke fun at those who opposed them. In 1850, the *Princeton Magazine* published a lengthy, humorous complaint against railroads. "My privacy has been invaded by the Railroad Company," the narrator brooded, "coming through my premises and cutting asunder my barn from my house, turning my back into my front, and setting fire to two haystacks by sparks from the locomotive." The author felt there was nowhere he could retreat on his property without being looked at by "curious passengers." He had hoped that his home would be a peaceful retreat, but instead, "all the world is passing every few hours, peering into my windows and scanning my petty garden, counting the hen-coops, the pieces in my laundry, and the very dishes on my frugal board, and ogling my respectable but too inquisitive wife and daughters, who have never been able to satiate their curiosity in regard to this intrusive wonder, nor to abide at any in-door work from the time they hear the sound of the cars." The railroad was literally tearing his home asunder, and figuratively did so as it decreased his putative control over the women in his life. "Home is no home, in sight of a railway," the author lamented. His chief desire was to be left alone. But the railroad made this impossible; it "laid their secrets open to the day."[49]

Steam transit, then, quickly became fodder for antebellum humorists. Early jokes expressed wonder at the remarkable properties of steam. Terminology used in steam transit lent itself to multiple, painful puns. Humorists were not afraid to examine the darker aspects of travel by alluding to accidents, and employees and passengers alike were ready targets for the humorists' or cartoonists' pens. In all, this humor demonstrates the broad reach of steam transit into American cultural life in both positive and negative ways. Riding steam may have been concerning due to dangers, but humor produced a way to dispense with that danger—brush it off with a joke. When humor poked fun at rubes, like the Georgia greenhorn who thought trains could fly, it solidified in the reader's mind that *they* were savvy users of technology, and not bumpkins. Here, humor could be provide cultural reassurance—the world was moving ahead, and the reader was moving along with it.

Music

Steam transportation brought with it its own unique set of sounds. Steam engines belched and screeched, and bells rang out to warn passersby of oncoming trains. This world of sound soon translated to the world of composed music. The arrival of steam transit inspired composers. Some composers created special marches or other pieces of music commemorating the creation of a new railroad. Other composers attempted to imitate steam through the sounds and rhythm of their own compositions. Composers did this both in sheet music written for piano that people could enjoy in their own homes and in music written and performed by large orchestras. As the antebellum era progressed, songs with lyrics praised the Underground Railroad, which used railroad imagery to discuss freedom. A rich body of composition emerged in the antebellum era, commemorating or imitating steam transit. I have identified sixty pieces of sheet music for piano with railroads or steamboats as an explicit theme (or with artwork featuring steam transit) published in North America and composed between 1828 and 1860, making it a regular topic throughout the antebellum era.[50] Pianos dominated private music-making during the antebellum era, and therefore steam transit was fully part of this musical scene.[51] As historian Billy Coleman has noted, "most of the music that early Americans heard was the music they made themselves"—music was a "participatory sport."[52] Thus, sheet music to be used in private homes was a critical way in which Americans experienced music in the antebellum era, and examples of music inspired by steam transit or simply featuring it as an illustration abounded.

At least four different pieces of music were written to celebrate the launch of the Baltimore and Ohio Railroad in 1828. The "Carrollton March"[53] was "performed at the ceremony of commencing the Baltimore & Ohio Railroad" and dedicated to Charles Carroll of Carrollton, a signer of the Declaration of Independence present at the groundbreaking of the Baltimore and Ohio Railroad.[54] The "Railroad March"[55] was dedicated to the directors of the Baltimore and Ohio Railroad and its sheet music cover featured a drawing of an engine pulling a train. The "Railroad Quick Step"[56] was a much shorter piece of music and was published with another piece dedicated to Carrollton on the same page. Finally, "The Rail Road"[57] was an arrangement for piano of several popular songs of the day. But the cover of its sheet music featured a remarkable illustration. The illustration depicts a location near Baltimore, with that city's Washington Monument visible on the left. At center, a carriage labeled "Ohio Velocipede" holds four passengers. In the foreground, a man and a woman call out to the passengers. The woman asks that they carry a message to her aunt that she will "drink tea with her in Cincinnati tomorrow evening and bring the new bonnet." The man requests that the passengers not forget to put his letter in the mail once they reach Wheeling, so that it "may get to N. Orleans the next day." Both of these comments allude to the power of steam transit to accelerate travel.

Eating successive meals in different cities was a common trope, demonstrating the distance that could be achieved in a single day. Both of their comments also hint at the link between railroad and steamboats, since finishing the journey to Cincinnati or New Orleans would require transportation beyond what the train could provide. Interestingly, we may surmise that the artist had not yet seen a train or did not wish to draw one. The engine, represented by a smokestack, is hidden behind a bush, and the car with passengers does not appear to be attached to the engine, thus having no way to move forward.[58] The artist's uncertainty illustrates the novelty of the technology and, perhaps, his own naïveté. The characters' commentary shows the excitement about the possibilities that steam transit brought forth (fig. 3.3).

Other musical tributes around the railroad followed.[59] In 1845, J. Guignard's "Alsacian Rail Road Gallops" appeared, and here we see a composer trying more explicitly to build the sound of the railroad into a piano composition.[60] The opening section consists of rapid motion in the left hand while the chords crawl upward in the right hand. This builds in tempo and volume, and when the tempo reaches "allegro," the text in the music reads, "LOOK OUT FOR THE LOCOMOTIVE." The passage culminates in a chromatic passage with the accompanying text which reads, "Smoke and hissing of the locomotive." On the sheet music itself, a drawing of a long chain of cars attached to an engine is integrated throughout all the lines on the first page of the sheet music, further tying the music to a representation of the machine.

Compositions—both to railroads and to steamboats—continued to come throughout the antebellum era. Ships were commemorated in the "Empire Quick Step"[61] and the "Knickerbocker Schottische."[62] Individual railroad lines or shipping companies also earned their own songs, such as the "Harnden's Express Line Gallopade and Trio,"[63] the "New Orleans and Great Northern Railroad Polka,"[64] or the "Erie Railroad Polka."[65] The "Fast Line Gallop,"

Figure 3.3. This cover of an early piece of sheet music demonstrates both the hope that railroads presented (through the optimistic comments of the onlookers) and perhaps some uncertainty about what a train looks like (the engine discretely hidden behind a bush). C. Meineke, "The Rail Road" (Baltimore: John Cole, 1828). *The Library Company of Philadelphia.*

dedicated to the president and directors of the Pennsylvania Central Railroad, featured on its cover a map of the railroad line and connecting services, charts giving distances between different cities on the line, and a full-color picture of a railroad traversing a bridge while people look on from a hilltop.[66] The "Chatawa Pic Nic Mazurka"[67] was apparently written in honor of a picnic site reachable by the New Orleans, Jackson, and Great Northern Railroad. The cover of the sheet music showed a train in the distance, with the tracks coming out directly at the viewer.

Smaller amounts of music featured lyrics. "Clear the Way!" celebrated the effort to build a railroad to the Pacific Ocean. The lyrics were a paean to "men of thought" and "men of action" who were pursuing this work. The lyrics hinted at the wondrous things to come with the completion of the railroad: "There's a fount about to stream, / There's a light about to beam."[68] In 1847, a song was published in favor of a particular railroad line in Maine, with the lyrics comparing the competing routes: "A voice from Bangor tells a tale / That makes our southern friends look pale—/Defeats their road, and puts it down,/And points the way to Lewiston." It closes with the rousing couplet: "Now clear the track for Androscoggin / The steam is up and we'll be jogging."[69] Music could also tell a humorous tale. The lyrics of "Ridin' in a Rail Road Keer,"[70] for example, narrate the story of a young couple who decide to get married having met on a train.

But some of the most powerful lyrics in the antebellum era came from songs that linked the physical railroad to the Underground Railroad. The song "Slaveholder's Lament" included the following reference to the Underground Railroad: "*Railroads* and stages through the wood, take '*things*' and make them '*men*.'"[71] This acknowledged the transformative power of the railroad to take people who were held as property and shuttle them to freedom. The song "Underground Rail Car"[72] featured a man who has escaped from slavery looking at a railroad. The lyrics announced that he was glad to "speed to day in the Underground Railcar." Likewise, "The Ghost of Uncle Tom" features railroad imagery.[73] When an enslaved person gets a "ticket" on the Underground Railroad, "Den his bosom's full of hope." The railroad is silent—"De Engine never whistles / And de Cars dey make no noise"—but the train can nevertheless "carry off" an enslaved person and his family to freedom.

Perhaps the best example of this genre was "Get Off the Track!" written in 1844 and performed by the famed antislavery Hutchinson singers. Before a single note of the piece sounded, anyone looking at the music would be able to discern the message. On the cover, a railroad car labeled "Immediate

Emancipation" is being pulled into a railroad station by an engine labeled "Liberator." In the distance, we see other trains crashing off of the track; only the "Liberator" can make it to its destination. The lyrics abound with railroad analogies, most of which signal that the train of freedom is moving expeditiously toward its destination. The lyrics gave a warning to politicians: "<u>Freight Trains</u>" of "<u>Votes</u> and <u>Ballot Boxes</u>" were coming toward them, indicating that they should heed the opinion of those voters. Another verse indicated why the other trains on the cover were failing: "Rail Roads to Emancipation/Cannot rest on <u>Clay</u> foundation/And the <u>tracks</u> of "<u>The Magician</u>"/Are but <u>Rail Roads</u> to perdition." By slighting Henry Clay and Martin Van Buren (nicknamed the "Little Magician"), the song urged listeners to stay on track for immediate emancipation. Other lyrics urged speed with railroad metaphors: "Wood up the fire!" "Roll it along!" and "Put on the steam!"[74] The overall theme was one of speed giving the concept of emancipation the inevitability of a train moving down the track. In his history of song in the United States, Charles Hamm wrote that "the effect of this song on antislavery audiences was electric."[75]

Steam transit was also invoked in orchestral music.[76] There appear to be several pieces that mimicked the sound of the railroad for comedic effect. In 1847, Caroline Healey Dall reported hearing one of these pieces performed in Boston: "The comic piece called the Railroad was very admirable."[77] The following year, a group called the Harmoneons performed in Cambridge, Massachusetts. They appeared in blackface for a portion of their program, and during that section they played the "Grand Oberture" of the song "De Rail Road Line."[78] In 1851, the same group had apparently kept this in their repertory, and while appearing in Worcester they closed the portion of their performance "As Ethiopians" (i.e., in blackface) with the "GRAND RAILROAD OVERTURE—A la Steyermarkische—Introducing imitations of Steam, Whistle, Speed, Locomotive, &c."[79]

The phrase in their advertisement—"a la Steyermarkische"—is a reference to the Steyermarkische Gesellschaft, a touring orchestra from Styria, in modern-day Austria. The leader of that group, Josef Gungl, wrote a piece called the "Rail Road Galop" which was apparently a popular composition on their tour.[80] In 1848, Joseph Mersman attended a concert by the Steyermarkische Gesellschaft in Cincinnati and found the overall performance to be "grand and beautiful, far superior to any orchestral Music ever heard in this City." The finale was Gungl's gallop, which Mersman termed "the most characteristic I have ever heard. All the various noises attending the starting of a train were most perfectly imitated. The piece instead of concluding with a general burst up, was brot to a decent halt, like a well regulated train of Cars should be."[81] The same orchestra included the "Railroad Gallop" at their concerts in Boston, New York, Washington, DC, and New Orleans, suggesting that it was a regular part of their repertoire.[82] A review of their performance in New York noted that the piece was "a beautiful similitude of an engine in motion" and "was received with reiterated cheers." When the group concluded its southern tour and was returning to New York, the New York *Herald* anticipated that the "Railroad Gallop" would "doubtless" be one of the pieces it would perform. Indeed, the performance of that piece "elicited the loud and continued plaudits of the entire audience."[83] Sheet music of the same piece was available for individual purchase at least by 1849, when it was advertised for sale by William Fischer in Washington, DC.[84]

The Steyermarkische players clearly left an impact because other groups took up the music, such as the Harmoneons mentioned earlier. The many references to "Railroad Gallop" in advertisements make it clear that other orchestral groups played the same music or similar pieces, bringing the sounds of the railroad to the audience. The Saxonia Orchestra played the Railroad Gallop in Concord, New Hampshire, in 1849, also using it to close its program.[85] One source claims that the Germania Musical Society performed a piece that featured a "little mock steam-engine . . . scooting about."[86] While historians are skeptical of the mock steam engine's abilities, there is no doubt that the Germania Musical Society incorporated the sounds of the railroad into its performances and advertised that they did so. In 1850, the Germania Musical Society performed in Worcester, Massachusetts, and ended its program with a "MUSICAL PANORAMA of Broadway, N.Y." The final item in the medley was "A ride to Harlem—Rail Road Gallop."[87]

Other groups played similar pieces throughout the remainder of the antebellum era. Public concerts in the United States during the late antebellum era were "utterly dependent . . . on large amounts of press attention and 'puffery' to generate curiosity," and the advertising language around steam transit-related pieces on the program reflects the effort to drum up excitement.[88] In 1855, a blackface group performed a piece described as "The Grand Excursion Party on the Express Train—the Station!—the Speed!!—the Explosion and Comical Denouement!!!"[89] In 1856, Warden's Renowned Opera Troupe performed "Railroad Gallope & Terrific Explosion, invented by Mr. Warden."[90] This suggests perhaps another piece of the same title. Another group, the King's Æolieans, performed in Worcester, Massachusetts, in 1858, and concluded its performance with "RAILROAD EXPLOSION! Giving imitations of the Steam, the Start, and the Stop—arrival at Squam Dunk Station—all aboard—can't get aboard—get a plank—Cow on the track—speed of the train, 1000 miles an hour—GREAT EXCITEMENT, and TERRIFIC EXPLOSION!"[91] Based on the advertisements for these performances, a piece of music imitating the railroad was often the last thing on the program. The orchestras sent their audiences off with a rousing, perhaps humorous, imitation of the railroad, and if Dall's and Mersman's reviews are any indication, audiences delighted in the effort by orchestras to emulate the railroad. As with the humorists who alluded to the disasters on railroads, these orchestral performances allowed audiences to think about the dangers at a safe remove. While disasters could be horrific, the musical invocation of steam travel could be enjoyable. Steam transit's movement into different parts of culture helped build the audience's comfort and familiarity with the technology.

Images

During the antebellum era, visual imagery increased dramatically in the United States.[92] Images of steam transit, ever popular during this period, were entirely a part of this broader growth. For example, steam transit featured in visual metaphors. In 1848, a broadside entitled "A Correct Chart of Salt River" mocked the problems that besieged Democrats during that election year. Candidate Lewis Cass is represented by a steamboat. Such a boat would seem to be an appropriate way to travel against the current, as the drawing depicts. But a closer look at the route spells trouble. The boat is marked "free trade," a reference to Cass's opposition to high tariffs. Cass's steamboat is about to face a series of disasters: Noise and Confusion Shoals and the Lake of Oblivion. By contrast, the (protectionist) Tariff of 1842 puffs ahead on a train, taking the direct route to Washington. Train tracks lead directly to the destination, avoiding the complex geography that awaits the steamboat. In this broadside, we see both types of transport recruited to make a political message: Cass's free trade was about to face treacherous waters, while the tariff has a smooth route to the capital city (fig. 3.4).[93]

A cartoon before the 1860 election featured Republican presidential candidate Abraham Lincoln and his running mate Hannibal Hamlin on a train labeled "equal rights." On the track before them was a wagon labeled "Democratic platform" being pulled in two opposite directions by the two Democratic standard-bearers: Stephen Douglas (with vice presidential candidate Herschel Johnson) and John Breckinridge (with vice presidential candidate

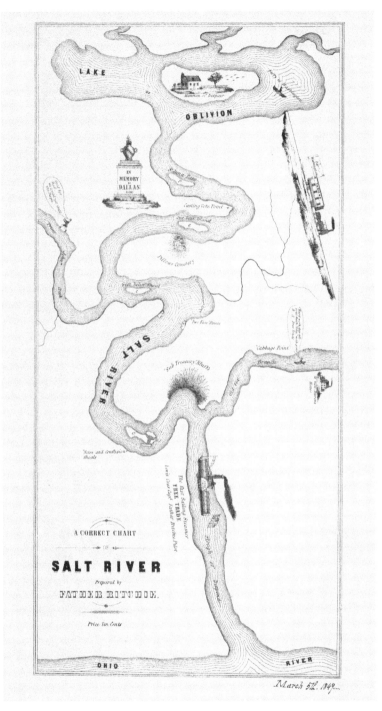

Figure 3.4. Different types of transportation are used to create a political message in this broadside from 1848. *The Library Company of Philadelphia.*

Joseph Lane). The cartoon's caption promises the "prospect of a smash up" as Lincoln's train comes barreling down the track toward the stuck "Democratic platform." Lincoln calls out, "Clear the track!" while Hamlin adds the familiar phrase "Look out for the engine, when the bell rings!" The trope of the engine approaching and the warning of the bell would have been familiar to all readers by this time and therefore would have emphasized the political message that it was time to clear the way for Lincoln's candidacy (fig. 3.5).

Images of transit were also frequently on currency. The United States did not have a single national currency prior to the Civil War, and so the plethora of banknotes and other financial paper created abundant opportunities for illustrations of steam transportation. Sometimes, a railroad or steamboat was almost unnoticeable in the background, a quiet reminder of its role in the economy. In others, railroads or steamboats were featured prominently, with goods being loaded or unloaded. In still others, railroads or steamboats were the center of attention, steaming across the middle of the banknote.

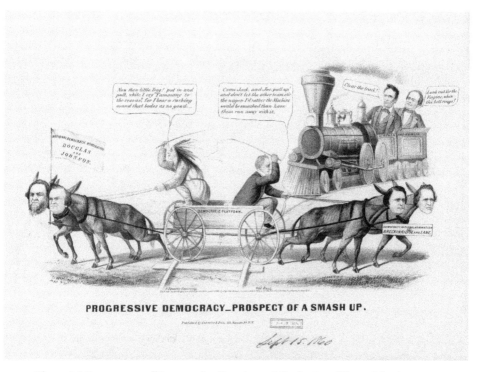

Figure 3.5. Democrats pull in opposite directions while the Republican ticket bears down the track in this political cartoon from 1860. *Library of Congress, Prints and Photographs Division.*

Modern transportation could also communicate a message about progress on the face of a note. On one $5 note, a steamboat cut through the water while nearby a Native American man attempted to tame a horse. Economic boosters and others paired the modern transportation with the inevitable eradication of native populations to draw a contrast of the fate of both peoples. Native Americans stood no chance when faced with the presumed superiority of white civilization, in this view. The image on the currency reinforced this interpretation—the Native American man and the horse were no match for the superiority of steam and progress (fig. 3.6). And one note from South Carolina explicitly showcased the southern economy through its illustration of the steam and enslaved power that undergirded it. Enslaved people, under the watchful eye of their master, load bales of cotton onto a wagon. A railroad filled with cotton bales pulls up to the edge of the water. A steamboat arrives to take the goods to their next destination (fig. 3.7). The image illustrates the power of intermodal transportation, widening the network for southern products, and underlines the racial hierarchy of the society. Southern businessmen and planters literally carried this reminder with them whenever they had this currency in their pockets.

Images of steam transit abounded in other contexts as well. In the mid-1850s, printmakers Currier and Ives "led the charge to celebrate the railroad and its contribution to westward expansion," producing series of prints on railroad themes that continued after the Civil War.[94] Glassmakers created flasks with "railroad themes" as early as the 1830s.[95] And artwork about the Mississippi River often featured steamboats. One spectacular example was

Figure 3.6. Five-dollar bill from the Bank of Chippeway, 1838. Catalog record 517150. *Courtesy, American Antiquarian Society.*

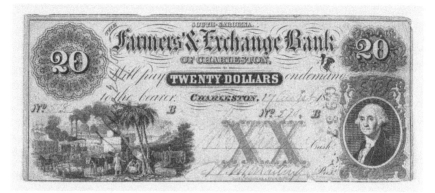

Figure 3.7. Twenty-dollar bill from the Farmers' and Exchange Bank of Charleston, 1853. Catalog record 517150. *Courtesy, American Antiquarian Society.*

the panorama, a massive rolled artwork that would unspool before an audience to give them the impression of moving on a journey. In the early 1840s, John Banvard traveled up and down the Mississippi River as he made sketches for a mammoth panorama of about 425 yards in length. The panorama excited audiences in Louisville, where Banvard did his painting, and he soon took it on tour elsewhere in the United States. The presentation of the panorama took several hours and was accompanied by Banvard's narration and live music. When he exhibited in Boston, special trains were run to bring people in from elsewhere in Massachusetts—steam transit thus facilitating an audience for steam transit rendered in art.[96] Such exhibitions continued throughout the antebellum era. A newspaper review of one such panorama in 1848 noted that it featured both "a steamer upset by 'snags,' and left an useless log upon the waters," and "a well-freighted vessel" showing the "wonderful prodigality of steam and human life."[97] The panoramic artwork was viewed as true-to-life, and it attracted attention in part because of the inclusion of steam. The artist featured both steam's triumph and steam being laid low by a snag, demonstrating the abilities and limitations of the technology.

Another panorama in 1848 included a dramatic image of a steamboat accident. A review of the panorama noted that it included "the burning steamboat, the pilot burnt at the wheel, the captain tearing the planks off the upper deck, the yawl upsetting, and females perishing," calling it "a sublime and terrible scene." Another reel of that same panorama had a fifty-foot-long longitudinal section of a steamboat, which featured "a correct representation of the pilot house, ladies' cabin, social hall, and the main saloon." Moreover, the panorama demonstrated proper activity on a steamboat: "The lady passengers are seen sitting at the table, while gentlemen who have no ladies under their care remain standing until the steward rings the bell, thus always securing seats for the ladies, no matter how great the crowd." The image even depicted the poorer passengers on board and a sense of how the entire machine was powered: "Below the cabins are seen the boilers, and the whole arrangement of a high-pressure engine with the accommodations for deck passengers; the whole being a correct view of the interior of the steamer Magnolia." This was followed by a representation of that same boat in a race with another steamboat.[98] This massive, mobile illustration of life on the river could show people who had never traveled west the operation of a western steamboat. Steamboats captured people's imagination, and panoramas such as these extended the steamboat's cultural reach.

While panoramas were massive and attracted large crowds, illustrations that took transportation as their subject existed in smaller scale as well. The satirical lithograph "Map of the Open Country of Woman's Heart" demonstrates how a map could be parodied to show the creator's viewpoint of women. It features a number of different regions or areas. The "Land of Selfishness," which includes the "City of Moi-meme [Myself]," is connected by the river "Indulgence" to the "Sea of Wealth." That river features "steamboat communication with Moimeme." A railroad starts in the "Land of Love of Admiration" and then travels to the "Land of Coquetry" before passing through "Jilting Corner" and the "Town of Lady's Privilege" and ending in the "Land of Oblivion." This railroad, according to the map, does "distance performed with incredible speed." Intriguingly, the areas of the map that presumably are being lauded by the creator—the "City and District of Love" in the "Region of Sentiment," which also consists of "Patience," "Good Sense," and "Enthusiasm," among others—are not served by railroads at all but by a series of canals. The map seems to hint that while the route to oblivion was fast and moved through unappealing areas, the parts of the heart worth cultivating offered slower traveling, perhaps signifying the need for good judgment and caution (fig. 3.8).

Travel Guides

Steam transit created its own genre of literature: travel guidebooks sprang up to inform potential travelers about the railroad and steamboat routes and what travelers would see along the way. Guides would help travelers avoid mistakes or relieve them of relying on strangers for information. Some of these

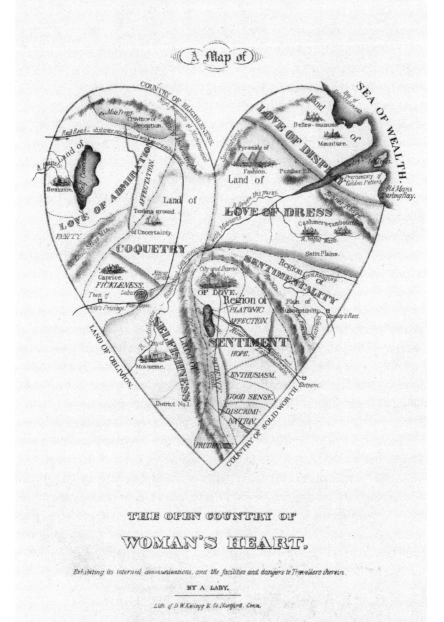

Figure 3.8. Many different types of transit are available through the complex landscape of a "woman's heart." Catalog record 150158. *Courtesy, American Antiquarian Society.*

guidebooks were purely tables of mileage, showing the distance between different stops along various routes.[99] Other guides provided mileage and linked together different routes to demonstrate how to get across the country.[100] Many other guides were highly descriptive. They painted a written picture of the route itself, transporting the reader into the train or steamboat so that they could "see" in their imagination what the route would be like. Others included dozens of drawings of the scenery along the way. In this way, the reader could envision their travel without ever leaving the parlor. Or someone for whom travel was not possible could likewise envision what it would be like to travel across the most distant reaches of the United States. These publications thus contributed to the commodification of travel. In the steam era, travel was no longer simply a matter of putting on one's cap and going for a walk or jumping on a horse. Rather, travel was something that could be purchased, in the form of a ticket; or, in this instance, armchair travel, in the form of guidebooks. These guides aided in that commodification by assisting in the description and "selling" of routes.[101]

With this wide variety of guides, travelers could purchase different degrees of information—they had a choice about what information they wanted to receive and how much detail they wanted to learn about given routes. Americans in the early nineteenth century used these guides to plot out their journeys. In 1849, *Horn's Railroad Gazette* urged travelers to use guides whenever possible: "Do you know precisely what route you have taken? Are you quite certain that you have made no mistake? There was another train which left the starting-place at the same time with yours. An error here would have been *easy*. Take out your Railroad Gazette and see by the landmarks that you are right, or if wrong you may perhaps get off at a way station, and thus avoid a failure in the important business which you have undertaken."[102] The guide thus underlined that travelers were no longer responsible for creating their own routes from scratch but could depend on the guides to show them the route.

Guides could reveal much about the landscape through which travelers would move. In 1836, Disturnell's guide gave an overview of the Hudson River and the times that steamboats departed there. The guide noted, "On leaving the wharf the view from the water is grand and imposing" and proceeded to describe the view in detail. The book continued to describe other towns along the way: readers learned that Hudson, New York, had "a population of about 6000 inhabitants," the wealthiest of whom worked in whale fishing.[103] Guides also commented on local oddities: an 1852 book on routes in New Hampshire informed travelers, "Near the summit of Red Hill there resides an old lady by the name of Cook. She has lived here many years; and in the summer of 1851 we found her surrounded by a deaf son and daughter (the former 60 years of age), besides several grand-children and great-grand-children."[104] Individual guides sprang up for specific states, such as Michigan or Ohio.[105] Other guides were devoted to specific lines, such as the Hudson River Railroad or the Pennsylvania Railroad.[106]

Guidebooks prided themselves on their illustrations, giving visual cues to travelers to watch for as they went along the route. In 1854, a guide covering a railroad in Pennsylvania boasted "that over seventeen hundred dollars have been expended for pictorial illustrations, some of which we can point to as fair specimens of the art."[107] In 1856, a guide for the Philadelphia, Wilmington and Baltimore Railroad opened with a nod to the illustrations by printing a humorous sketch that purported to show "the perils of our artist, in search of the picturesque, during the prevalence of a fog." The picture shows the train on the tracks, having hit a hand-car that held the artist. The result was that he was sent scrambling, with his equipment flying through the air.[108]

One of the dominant guidebooks of the period was Appleton's, becoming "essential" for railroads and advertisers alike.[109] The editors hoped their work would find that readers with "the eye of patriotism will here see portrayed those mighty works, whether completed or in progress, that are bringing the most distant parts of the Union into neighborhood; and which, by blending into one the interests of the East and the West, the North and the South, are creating an additional guarantee for the repose and permanence of our great confederacy." Acknowledging that things were always changing, the editors wrote, "We pledge ourselves to keep it up to its present standard of usefulness by thorough periodical *revisions*."[110] The guidebook, then, was not just a tool for the traveler but a tool for learning about the country at a time of sectional tension.

Appleton's editors thus alluded to a useful part of these guides: descriptions of the land, telling the traveler what to expect when along the route. If the traveler read the guide in advance of the trip, it could build anticipation for the moment of seeing the most glorious sites. If the traveler kept the book with her in the train, she could read along as the scenery whizzed by, learning a great deal about her surroundings. An 1851 guide to the New York and Erie Railroad gave a detailed description of the route and scores of drawings showing travelers what to expect (or showing them what they may not be able to see, depending on where they are sitting). Regarding the station of Summit, the book urged the traveler to "bribe the engineer for the privilege of one of his windows" and thereby see the extraordinary landscape at that station.[111] Other guides also placed an emphasis on what travelers could see—or should not bother seeing. That same year, the *Illustrated American News* published a travel guide in which some of the scenery along the road was described as "common-place, and the traveler coming from the West, who is transferred to these cars after dusk, may assure himself that he is not deprived of any sight worth seeing."

Later in the route, the reader was alerted that seeing the workshops of the railroad was an interesting sight, but by no means should a passenger "risk the loss of a good seat in the cars" in order to view them. Thus, the guide provided specific directions on where to sit in order to get the best views but also assured the reader when what was being missed was not worth the loss of seat. Speaking of a particular bridge, the author urged the reader to *"be on the alert to seize the actual moment of crossing the bridge,* for the train flies so fast, that he would lose the glance into the gulf beneath in three winks of an eyelid. Seize the moment, friend and fellow tourist, and do not recoil. (Remember, you must be on the right side of the cars.)"[112] Readers were urged to get in the proper position, have courage, look down, and witness themselves riding above the tops of trees.

The authors of these guides were cognizant of the changes that steam transit wrought on American life. In 1852, William Guild published a guide to New York and the White Mountains. It opened with a reflection on how railroads had changed travel. Prior to the availability of easy travel, the mountains were not considered as a destination, for "the want of an easy method of reaching their neighborhood. . . . But all this is now changed. The comfortable railroad car now flies daily to the very base of these stupendous heights, upon all sides."[113] A guide to Ohio railroads in 1854 opened with an appeal to history and progress. Referring to Cincinnati, the guide opined, "In truth, all that you see in this metropolis of the Ohio Valley, is the result of only a half a century of *hard work*. . . . Soon the Indian disappeared; . . . the town grew up, at first only a dirty village, and now a great and prosperous city, full of art, commerce, and wealth. No where else can so entire a transformation, accomplished in so short a time, be found."

The guide continued to reflect on changes to society: "What a revolution! But the revolution is not in the gain of time only, nor even money. The great change is in society. Thousands meet now where tens could meet twenty years since. Look through these cars, and you see around, men, women and children going to see friends, or transact business, or seeking pleasure, where they would not have dreamed of going a few years since. . . . Where will this stop?

No where, till this earth is inhabited by one family, dwelling together in peace and unity."[114] An 1857 guidebook also naturalized the conquest of the land and the death of the Native Americans when it noted that some "broken and sterile" land was "once the hunting ground of the Indians, that unfortunate race which was disappeared from this land, leaving no monuments to mark their familiar spots, or to perpetuate their memory." The writer left unanswered the question of how that disappearance took place, but rather saw the death of the Native Americans as inevitable: "But such was the destiny which the God of nations had fixed for them, and His purposes are beyond our ken."[115] In short, the guides did not simply provide descriptions of the routes but were fully aware of the cultural import of steam transit.

From lists of mileage to philosophical rumination on the fate of nations, these guidebooks offered much to the American traveler. They represented a new type of literature that grew significantly during the antebellum era, one that was the direct result of new modes of steam transit. At the same time, these guidebooks were but a part of a larger web of steam transit in the broader culture. To be sure, it took some familiarity with technology for the initial metaphors, jokes, or songs to have meaning, but the spread of cultural markers about steam transit outpaced the actual physical presence of steamboats and railroads. Even miles from the nearest station or wharf, Americans could read or hear cultural expressions that invoked steam transit's characteristics. Aspects of steam transit lent themselves easily to metaphors in public and private writing. Humorists took steam transit as a subject, teasing rubes new to travel, expressing astonishment at steam's power, and darkly joking about steam's destructive force. Music composers celebrated steam transit's arrival, and they even attempted to employ steam transit's sonic markers in their compositions. For people who never rode a train or steamboat, these cultural products offered additional entries into the world of steam: in a letter from a friend, a cartoon in an almanac, an orchestral concert, or a guidebook read in the comfort of the home. Steam transit was a physical thing, but it quickly exploded in the cultural realm to be a conduit for talking about a plethora of topics: time, power, movement, relations between men and women, and a host of other subjects. Steam's power was not just in transportation, but it traveled much farther through its impact on the arts. Indeed, steam's cultural resonance was fundamental to the naturalization process, and through this means even those distant from steam could become engaged in the Transportation Revolution.

CHAPTER FOUR

Religion

The expansion of steam transit in the United States arose around the same time as the Second Great Awakening—a time of remarkable religious ferment. As historian Daniel Walker Howe has noted, religious thought in the Second Great Awakening was not simply confined to Sunday mornings. Rather, the awakening could be seen throughout society. It had an impact on "literature, politics, educational institutions, popular culture, social reforms, dietary reforms, utopian experiments, child-rearing practices, and relationships between the sexes."[1] Religion in this period was dynamic, incorporating the changing world around it.[2] Therefore, it should not be surprising that technology and steam transit were also considered subjects worthy of exploration by religious thinkers. Science and religion were not "independent entities" in the early nineteenth century.[3] Confronted with rapidly changing technology, American religious leaders did not condemn it outright or ignore it but rather wrestled with the implications of changing technology for God's creation. An examination of how religious thinkers addressed steam shows that they did not see a boundary between religion and technology but thoughtfully considered how one could influence the other.

Most of the religious writers who commented on antebellum steam transit were Protestant, and there do not appear to have been significant denominational differences within that group. For many Protestant writers and thinkers, the expansion of steam transit was a positive good. For them, the presence of beneficial technology in the United States demonstrated God's favor on the United States. More generally, these thinkers argued that technological advancement occurred because God had endowed humans with a "creative spark." Steam transit may have been created with improved commerce in mind, but trains and steamboats could also carry the Gospel rapidly around the country and thereby allow religion to reach even more people. Thus, steam transit was a gift from God, to be used wisely in carrying out God's will.[4] In that respect, religious thinkers were in a long line of Americans who used technology to spread the Gospel—radio, television, and the internet would likewise be used by religious groups in the ensuing years. In the antebellum era, this particular strain of religious thought, supportive of steam transit, also contained within it a current of anti-Catholicism. These preachers and writers looked at the apparently slower development of railroads in Catholic countries and the alleged denunciation of railroads by Pope Gregory XVI and saw both as signs of Protestant superiority.

There were some Protestant writers who looked at steam transit with a more skeptical eye. Almost none opposed steam transit outright, but there were many who wanted Americans to think carefully about the moral implications of steam transit. Some preachers feared that the overwhelming pursuit of wealth, of which transportation was a part, was distracting Americans from their spiritual development. Troublingly, the fast-moving age was leading people from the path of religious improvement. Moreover, many steamboats and railroads operated on the Sabbath, leading to many arguments about the moral effect on passengers and employees alike. Thus, there was a tension in religious thought with respect to modern technology. Steam transit both made possible the expansion of the Gospel's reach and tempted people away from the Gospel's lessons.[5]

Although ministers would take up steam transit as a topic on their own or might be invited to offer a prayer or sermon when a new railroad was launched, one common moment for ministers to reflect on steam transit is when they delivered eulogies for people killed in railroad or steamboat accidents. Americans looked to religious leaders for guidance in the wake of such horrific events. Steam accidents seemed to strike at random, without respect to the age, race, gender, or class of the victim. Preachers delivered sermons and eulogies that attempted to make sense of how transportation that offered so much promise and opportunity could also deal in death. Ministers often took the opportunity to underline for their flock the importance of living a good life and being prepared for death at any moment. Steam transit, then, provided the perfect opportunity to teach a lesson about the importance of spiritual preparation prior to death.

Religion played a major role in antebellum American life; therefore, any consideration of the cultural impact of steam transit would be incomplete without a look at religion. As we will see in this chapter, religious leaders had different reactions to steam transit; again, the common experience was not necessarily unifying. Indeed, three major Protestant denominations underwent schism in the antebellum era.[6] Protestant ministers may have united in believing that steam transit could help spread their message, but the precise content of that message remained up for debate.

Steam Transit as God's Gift

When surveying the response of antebellum Protestant writers who commented on steam transit, it is clear that the majority of the responses were quite positive. There were several themes that ran through these responses. First, these religious leaders saw the development of steam transit in the United States as a sign of God's favor toward the country. Second, steam transit was a sign of civilization or progress. For these authors, it was logical that Christianity and civilization went hand-in-hand; therefore, they saw steam transit as beneficial. Third, steam transit could assist Christianity in spreading the Gospel. It was quite literally the fastest way to travel and should be pressed into service to bring nonbelievers into God's fold. From the printing press of Gutenberg to twenty-first-century televangelists, religious leaders have often accommodated technology if they could use it to spread their message. In the antebellum era, steam transit was no different. Ministers rarely focused on one of these advantages but often wove them together in their sermons or writings. They served as interlocking and self-reinforcing justifications for steam transit: it showed God's favor on the United States, was a partner with Christianity in civilization, and could help spread the Gospel.

Some ministers explicitly preached about steam transit. In 1846, Horace Bushnell gave a sermon titled "The Day of Roads." He opined that if a people "have no roads, they are savages; for the Road is a creation of man and a type of civilized society." Having suggested that roads of any type reflected civilization, Bushnell brought the message to contemporary times, observing that "horses ... of iron" had replaced "horses of flesh." Bushnell saw in railroads "the new age breaking through the old," and he christened this new era the "age of Roads." After underlining the importance of roads, Bushnell asked if "our age of Roads has some holy purpose of God fulfilling, in its social revolutions, ... the coming reign of Christ on earth." In so doing, he lifted thoughts about roads from mere economics to a higher purpose. Bushnell answered his own question in the affirmative. God's plan in promoting transportation was to bring the world into unity. Bushnell even located biblical justification for this view; in the book of Isaiah there was a prophecy that

Protestants saw as a prediction of the railroad's value: "The crooked shall be made straight, and the rough ways shall be made smooth."[7] Bushnell's address typifies antebellum sermons that linked progress, steam transportation, and the advancement of Christianity.

Another minister, S. C. Aiken, combined many of these themes in his 1851 address "Moral View of Railroads," which he preached at the opening of the Cincinnati, Columbus and Cleveland Railroad. As did Bushnell, Aiken began with the importance of progress. "Point me to a country where there are no roads," he said, "and I will point you to one where all things are stagnant—where there is . . . no learning, except the scholastic and unprofitable. A road is a sign of motion and progress—a sign the people are living and not dead." While the railroad was critical for economic advancement, its value did not stop there. Rather, God had a specific plan for railroads. They were "intended by Providence to act upon religion and education." The railroad was thus intended to be a tool for the expansion of Christianity.[8] Like Bushnell, Aiken found a biblical reference that supported his views. In the Old Testament, Nahum 2:4 prophesied that "chariots shall rage in the streets" and they shall "run like the lightnings," which must have seemed to Aiken like the closest description to a railroad that was available in the Bible. Newspapers across the country noted Aiken's sermon, including his apt biblical reference.[9]

Finally, in 1850, Joseph Wilson delivered a sermon in favor of the Pacific Railroad. He, too, attached railroads to civilization: "The construction of a great national highway is . . . an evidence of a high degree of civilization." And civilization and religion went hand-in-hand. The purposes of roads were to "facilitate social intercourse and to extend the blessing of civilization and religion." He believed that the railroad, although designed for commerce, would invariably carry religion along at the same time. The end result would be the expansion of the Gospel not just across North America but across the Pacific Ocean and into the Asia. With the construction of a railroad to the Pacific, "the elements of christian civilization will . . . find a home on the genial shores of Asia; thus bringing the most distant nations of the earth into intimate commercial and social relations, and illuming the dark realms of Paganism by the light of heavenly truth." He concluded that railroads were no less than "the 'Acts of the Apostles' of our civilization."[10] Technology would help to spread the word of the Bible, just as the apostles had spread the word centuries prior to Wilson's talk.

Such dedicated sermons make explicit the links that Protestant ministers saw among technology, civilization, and Christianity and how they were willing to adapt to this technology by addressing it in their preaching. Far more common than a dedicated sermon, however, was for writers to make similar points in larger works not dedicated to transit. In 1827, a writer in the *Christian Visitant* argued that "the spirit of improvement is in full march, and directed to every object within its reach." The writer tied the improvements in religious education to improvements outside of the spiritual realm, such as "the growing attention to canals and railroads." For this author, then, improvements in spirituality and improvements in technology were linked.[11] The following year, an anonymous writer wrote that "speedy communication" had a "definite object" which "requir[ed] a greater facility and rapidity of transit." In that author's view, that chief object was Jesus Christ, and so anything that aided Christ's work would be beneficial.[12] Transportation could not just be for its own purpose; it had to support the growth of religion.

Alexander Bradford took the presence of steam transit in 1845 as a sign that all (rich and poor) could share in God's bounty. Looking at ancient culture, such as that of Egypt, Bradford found that the "monarchs lived in superb mansions" while the common man would only "sweat and toil for others." By contrast—and conveniently forgetting the enslaved people in the South—Bradford noted that in his own time, "we find knowledge diffused and the arts cultivated for the benefit of all.... The artificial rivers, the long lines of railroads, the ceaseless manufactory, the leviathan steamer, the printing press, these are not of right nor in fact for privileged classes, but the humblest of us may enjoy them and their products at a moderate cost." All public works were there to "bless with the best of God's bounties, each denizen of this great city."[13] This minister saw steam transit as integral to creating equality. Everyone could benefit from the technology. This equality was a gift from God, as was the technology that had delivered it. The reality in the United States was of course far short of the ideal, but Bradford believed that steam transit delivered on that promise.

In 1851, the Reverend P. D. Huntington called the railroad "a mighty bond... holding together all the most precious interests of civilized society,— those also which the Christian church both wants and watches." He argued that railroads necessarily brought freedom with them: "Motion is freedom. The steam-whistle is a trumpet of universal emancipation. The iron rail is a more effectual lecturer against every form of oppression than any of the anti-slavery agents." Huntington termed the engineer driving the train an "unconscious missionary," since railroads could bring Christianity everywhere, even if they were built for a different purpose. He concluded, "Over these iron paths God will roll in upon us an age of greater toleration, helpfulness, and love." He was confident that railroads would be the instrument of achieving Christianity's work: "We may confidently believe that the gospel *is* to be published over the round world, not by an angel, but by earnest men; and borne, not by celestial wings through the air, but by railroads and steamships over the land and over the sea."[14] Like others before him, Huntington linked civilization and technology in a way that would lead to expanding religion.

Ministers argued that even if technical improvements were started for purely monetary ends, Christians could appropriate them for their own ends. A writer to a Presbyterian newspaper in 1850 urged his readers onward: "If others travel with rail-road speed, must we still cling to the stage-coach?"[15] Steam was a proven technology, and Christians, too, could benefit from steam technology, even if the original intention of the steam transit was more worldly than divine. "Capitalists lay railroads, but along the iron track, Christianity and its literature, and its thousand appliances for good, travel and are diffused," Joseph Copp pointed out in 1858.[16] Christians could improve upon the wealth-seeking properties of railroads by using them for their own, holier ends.

Throughout the country during this period, steam transit was invoked as a metaphor in public and private correspondence alike. Religious authors were no different in this respect. In 1836, the *American Quarterly Register* counseled patience to those who worried that the work of converting people to Christianity was not going fast enough. The editor underlined that missionaries were "not constructing railroads, nor making a turnpike over a mountain. If they were, the business might be done with all speed." Instead, they were doing the much harder work of trying to change people's souls, which required time and patience.[17] In 1838, the *Colored American* used steamboats as a metaphor in a religious instruction. The writer told that he observed a man rushing to catch a steamboat. The man, however, was too late. The author used this example to illustrate the importance of timeliness in saving souls. "The poor man looked very sad, bit his lip and stamped his feet, but all would do no good, it was 'too late,'" the author observed. No matter what the man's reason for wanting to board the steamboat, he would not be able to board because it had already departed. "How many, my young friends, are too late about religion," the author wondered. "That most important of all things. We see so many who have put off religion till it is too late, that there is great reason to fear that many more will do the same thing."[18] Steam transit was known for requiring punctuality, so this lesson would have meaning to the reader.

Laypeople also considered how religion and transportation intersected. While traveling in South Carolina, Charlotte Hixon reported that she "had an interesting conversation with Mr. Parker respecting several passages of scripture. The prophesy, 'The crooked places shall be made straight, and the rough places smooth' he thinks is literally fulfilled in the laying out of Railroads, all over the country."[19] In the remarkable engineering works of man, some saw the fulfillment of a long-promised prophecy in Isaiah. If one believed in the connection, then the creation of railroads in the United States was not just a marvel but something ordained by God.

Steam could also be found in religious imagery. The 1857 *New York Comic Almanac* featured a visual pun with engines on two different widths of track and the caption "The Broad and Narrow Way." This was a reference to Matthew 7:13–14 (King James Version Bible): "Wide is the gate, and broad is the way, that leadeth to destruction," and "strait is the gate, and narrow is the way, which leadeth unto life."[20] This almanac took the familiar saying and turned it to refer to different gauges of railroad. Imagery had the potential to deliver much stronger messages as well. One such image was in support of the temperance movement, the movement to encourage abstinence from all drinking. Around 1841, the artist J. H. Knowlton drew a temperance image that featured a train of cars pulled by a wild boar. The car being pulled by the boar is labeled "New England Distillery" and is full of drunkards. Riding on top of the boar, the "engineer" holds a bell in one hand and a bottle in the other and calls out with the familiar phrase "Look out for the engine while the bell rings." On top of the car, the devil pumps a bellows which stokes a fire underneath a still. All around the train, men, women, and children stumble, while underneath the train men are crushed, as well as documents reading "Statutes," "Love," "Truth," and "Moral Reform." The artist here used a now-familiar technology to help make its point: alcohol was like an unstoppable railroad driven by the devil and leaving destruction in its wake (fig. 4.1). Fighting for temperance would require strong work indeed to fight off such a strong opponent.

For many Protestant religious writers, then, steam transit was a positive development, one that they applauded and welcomed into religious life. For a subset of these writers who looked favorably on steam, steam transit was also an example of Protestantism's superiority over Catholicism. Most of the evidence in this entire chapter comes from Protestant sources, and for some Protestant writers this anti-Catholic element was crucial. Protestant writers believed that Catholics opposed railroads in part due to the supposed attitudes of

Figure 4.1. Intemperance as a train crushing all else beneath it in this 1841 temperance image. Catalog record 468454. *Courtesy, American Antiquarian Society.*

Pope Gregory XVI (1831–1846). According to historian John Pollard, Gregory was "reputed to have said 'chemin de fer, chemin d'enfer' (roughly translated: 'the iron road is the road to hell')." Gregory was apparently worried that railroads bringing cheaper goods from Europe would wreck the local, rural economies of the Papal States. But in the eyes of anti-Catholics, the French pun was evidence that Catholics were opposed to technology. As we have seen, Protestant writers linked technology and the advancement of civilization to the advancement of Christianity, which made Gregory's attitude all the more problematic. Gregory's successor, Pope Pius IX (1846–1878), was reported to have "rather enjoyed a train ride" and was more supportive of constructing railroads.[21] For Protestant writers looking for additional ammunition against Catholics, however, Gregory's supposed attitude told them all they felt they needed to know.

Thus, for much of the antebellum period, Protestant writers saw the putative attitude of Catholics toward railroads as evidence of Protestant superiority. In 1837, the *Methodist Magazine* quoted a professor giving a lecture on education. He pointed out that the United States benefited from the locomotive engine, which was "that most brilliant gift of philosophy to man." He contrasted locomotive in the United States to transportation in Rio de Janeiro, where "the ordinary vehicles that meet the eye are ox-carts." The reason for this contrast, he concluded, is that the United States placed a high value on education. By contrast, the South Americans were "besotted in ignorance." The reason for this difference, in turn, was attributable to religion: "We have a ministry whose minds have been enlarged and improved by knowledge; they are shackled and oppressed by a multitude of ignorant and vicious Catholic priests."[22] This Protestant accused Catholics of keeping their parishioners in ignorance, and thus shunning railroads as well.

Other Protestants picked up on this theme. "Our fathers fled from the papacy," the Reverend H. P. Tappan wrote in 1848. The result, in his mind, was a distinction between the Catholic and Protestant worlds. "Their population is prostrated in ignorance, filth, and beggary," he wrote, "while ours have schools, books, newspapers, property, wealth, and freedom. They may have St. Peters and statues, but we have manufactures and railroads."[23] Once again, steam transit was a marker between the more civilized and advanced Protestant civilization and the less advanced Catholic. Protestant whites valued their independence, and they saw Catholics as dangerously dependent on the church in Rome.[24] More anti-Catholic thought came in an 1849 article. According to that author, thanks to the "intellectual light that is blazing over this continent," Americans used technology while Catholics remained in darkness. The author tied this explicitly to steam transit and its ability to carry information to every corner of the country: "The day of railroads is come—the day of steam—of the printing-press—of the magnetic telegraph. There is now hardly a backwoodsman who is not a traveller; there is hardly a log cabin beyond the mountains whose inmates are not made weekly acquainted with what is going on in the very ends of the earth."[25] Steam and other technologies brought education and enlightenment everywhere and people were better off because of it, in contrast to the Catholic countries, which remained ignorant.

Other writers hoped that the benefits of railroads would rub off on Catholics who happened to be in proximity. The *Well-Spring* reported on a "Railroad Jubilee" held in Boston. The author noted that there were "many hundreds of French *Catholic* Canadians" present at the celebration. This was significant, because when these Canadians traveled to Boston, "from the time they entered our territory, they every where saw the fruits of *Protestantism*, in the quiet, beautiful villages, with their churches and school-houses side by side.... Can such an exhibition of the fruits of Protestantism be witnessed without influences for *good*? We think not." For this writer, God's intent for the railroad celebration was, in part, to spread Protestantism's influence to Canada: "The easy and rapid mode of communication now completed, and which has just been celebrated, will soon make the Provinces familiar with all our benevolent and religious institutions."[26] The following year, the *Well-Spring* continued to argue that the railroad was a demonstration of the superiority of Protestantism. After noting that Spain had "no railroads, no canals, [and] no telegraphs," the writer posed the question: "But why does Spain linger so long in ancient darkness? Why is she, and why is Portugal, and Austria, so far in the rear of the world? There is but one answer. They are Papal countries, and Popery never advances but into thicker darkness."[27] For these Protestant authors, the rejection of steam transit was indicative of a larger rejection of knowledge within Catholicism.

Another anti-Catholic work in 1856 used a Catholic paper's discounting of railroads as a sign of how the Catholic church worked to keep its followers in ignorance: "How are the poor papists to understand it, Americans, when the priests keep them in ignorance, by shutting out the light of truth from their minds? The leading French journal of the 3rd of April, this year, speaking for the Romish church, says: 'Railroads are not a progress, telegraphs are an analogous invention; the freedom of industry is not progress; machines derange all agricultural labor; industrial discoveries are a sign of abasement, not of grandeur.'"[28] The meaning of this quotation was clear for Protestant writers, who latched onto the Catholic church's reputation for opposition to railroads and other indicators of progress as a sign of its backwardness. Catholicism was seen as changeless, and thus it was unable to adapt to or appreciate technological changes and their advantages.[29]

Spreading the Gospel under Steam Power

Religious writers praised the advantages of steam with high-flown rhetoric and weighty aspirations. But the uses of steam were not merely theoretical for antebellum religious leaders. Traveling on steam transit presented both formal and informal opportunities to discuss religion. Many accounts attest to transit being used in this way. In 1836, a writer to the *Sunday School Visiter* wrote that while traveling on a steamboat, he had spoken about his religious views, which had apparently come under attack from a deist. He "felt it to be my duty to defend the Christian religion" and hoped that in a public area such as a steamboat, a public refutation would also have an effect on other passengers. The deist "had publicly proclaimed the Bible to be *full* of contradictions" but declined to name any when handed a Bible. The author reported that he effectively quieted his rival, and the deist "remained silent upon that subject the rest of the trip, and treated me with all possible respect, notwithstanding the argument we had had."[30] Steam transit presented a large—and captive—audience for religious instruction, much as it had provided an audience for abolitionism.

There are other references to preaching on steam transit or using it as a site for moral reform. English missionary Joseph Wolff published a letter in 1837 in which he noted that he preached "most frequently three times a day, and sometimes in steamboats."[31] Another such missionary was mentioned in the *Colored American*, which reported on a "free man of color" who had built a steamboat that included a library and was used as a "temperance boat" on which the man would travel and promote teetotaling.[32] There were plenty of steamboats that carried alcohol as cargo, but this particular minister had repurposed a steamboat to advocate for temperance, a common religious cause in the antebellum era. In 1845, a newspaper item recorded that on a steamboat traveling from New York to Providence, the passengers requested a sermon on board. The captain agreed, and so "the spacious dining saloon was assigned for the services. After supper, invitation was given out, and a good congregation, including the captain and his lady, immediately assembled."[33] Here, the passengers welcomed an opportunity for religious instruction.

Religious missionaries saw steam transit explicitly as a way to spread their Gospel. In 1842, the board meeting of the American Tract Society contemplated that the work of teaching the Gospel could happen anywhere: "The whole land is a vast school. The rail car, the steamboat, the manufactory, the work-shop, and the farm-yard, the mines of the Schuylkill and of Galena, are all *schools*." The railroad made the constant circulation of information possible, its "unceasing clatter echoes among the hills all day, and his fiery train illumines our valleys at night." The desire for reading material was readily apparent: "Everything seems to conspire to arouse and excite the public mind, and reading it will have."[34] The printing press and steam transportation together made circulating material possible in ways that were never possible before. Christians, then, had to ensure that these tools were used to spread the Gospel.

Beyond spreading the Gospel, religious leaders also used transportation infrastructure to conduct religious services. In 1856, a report of the Presbyterian Church noted that missionary preachers were preaching in a variety of locations including "'a ball-room,' 'a distillery,' 'a barn,' 'an unfinished rail-road depot,'" and others.[35] The following year, the *National Era* wrote that a preacher in Illinois gave a sermon in a railroad car, with two railroad corporation presidents in attendance. In lieu of a church bell, "the congregation was called together by the bell of the locomotive."[36] The Baltimore *Sun* reported in 1857 that in Iowa City, Iowa, a "large population" had emerged around a railroad depot, and so a Sunday school was "held every Sabbath in a railroad car in that place."[37] And in Wisconsin in 1859, an "eccentric Episcopal clergyman" baptized a child while riding on the railroad. The minister "repeated the baptismal service from memory, and the conductor held the basin of water."[38] These examples illustrate how religious leaders used railroad infrastructure to deliver their religious messages. Sometimes it was inspired by the moment, and sometimes transportation infrastructure was used because it was the only thing available. For these Protestant religious leaders, a willingness to hold services or perform sacraments anywhere included the country's burgeoning transportation infrastructure.

Not only was steam transit used as a site for religious engagement, but the infrastructure could also be used to expand religious engagement. Although he did not preach on the train, Frederick Plummer recorded in his diary in 1850 how trains allowed him to travel around New England and preach.[39] In 1851, another writer reported to a religious newspaper that "in two years I have, on Sabbath mornings, visited about six hundred steamboats, and have presented personally not less than 20,000 individuals with tracts, and a very large proportion of them have been read."[40] The American Tract Society owned steamboats, assisting in its missionary work on the Mississippi River. Indeed, according to historian John Lardas Modern, "by the mid-nineteenth century the history of evangelicalism had become all but equivalent to the story of technological triumph." Evangelicals were anxious to use any tool that allowed them to expand the reach of the Gospel, and if steam transit brought about that end, then they embraced it enthusiastically. Such evangelicals accepted that engineering and scientific knowledge were not antithetical to religion but a necessary part of achieving their ends.[41] They had already accepted the printing press as a means of increasing the circulation of their ideas, and the addition of faster transportation merely extended what was already in process.[42]

In 1852, an H. M. Saxon reported in the *Age* how transportation networks could be helpful to religion in an article titled "Distribution of New Church

Books." After detailing the thousands of books and tracts he had distributed in dozens of cities, the author noted that "some were distributed in the country away from villages and cities, and some on cars and steamboats." He clearly used the transportation networks to do his work, and sometimes delays that vexed other travelers he deemed auspicious. When the railroad was "providentially detained" on a trip, he took the opportunity to speak to a stranger for "one to two hours." After this chance encounter, the stranger "became deeply interested in the Heavenly Writings, and may now be regarded as a receiver of the New Church Doctrines." Saxon further recounted another episode on a railroad, in which he "selected a man who I thought would buy a book, and took a seat by his side. I handed him a copy of Heaven and Hell, which he looked over with interest, and after a very pleasant conversation, he bought the book and said he would read it."[43] Steam transit offered opportunities to proselytize, which men like Saxon seized. Potential converts were boxed into the same car or salon as the person doing the proselytizing.

An etiquette guidebook with specific instructions for Christians underlined the importance of looking for such opportunities. The guide warned that when a Christian was "traveling as a man of business and passing over his route with the utmost speed, he is liable to be less attentive to the means of grace, and to relax his habits of watchfulness and prayer than when he is at home." The traveler was urged to "engage much of his attention in reading his Bible, or some other spiritual or intellectual book" unless he was able to "form an acquaintance with some pious or sober-minded person" instead. Christian travelers were also encouraged to gaze out upon the scenery, since it was impossible for a person to do so without "elevating his views of the wisdom, skill, power, and benevolence of the earth's great Architect and Supporter."

The book also emphasized that the "Christian traveller should always be watching for souls. Before setting out he should provide himself with Testaments, tracts, and other evangelical books." Thus armed, the Christian traveler would be ready to discuss religion with other passengers. Travelers were cautioned to "introduce the subject with unobtrusiveness, humility, and gentleness" and to "avoid as much as possible, sectarian views and every appearance of cant." In such a way, the Christian traveler could take advantage of the fact that "many travellers converse on the subject of religion more candidly and freely with a stranger, than with their most intimate friends."[44] Travel led some people to let down their guard and talk more freely. This guide urged Christians to be alert to such moments and embrace them for their opportunities for evangelism. In this way—with temporary acquaintances and the opportunity to meet many people—steam transit was perfectly suited to expanding the reach of the Gospel.

The advantages of steam transit, then, were not merely that it allowed for the spread of the Gospel by carting boxes of Bibles or tracts or religious newspapers around the country. Rather, the train and steamboat themselves could serve as locations for religious observance—impromptu or planned. Religion was an important part of early American life, and as such, the willingness of religious thinkers to discuss and interpret steam transit was therefore an important aspect in steam transit's normalization. Traveling in the United States in the 1840s, Scottish journalist Alexander Mackay noted that the design of American rail cars (different from that found in Europe) was "like a small church upon wheels. At either end was a door leading to a railed platform in the open air; from door to door stretched a narrow aisle, on either side of which was a row of seats, wanting only book-boards to make them look exactly like pews, each being capable of seating two reasonably sized persons."[45] The appearance to a church may have been accidental, but creative religious missionaries seized the opportunities provided by steam transit and its infrastructure.

The Temptations of Steam Transit

Not all religious thinkers saw steam transit as an unalloyed good. Some cautioned that it presented moral challenges that had to be met if it was to be amenable to Christian society. The most common cautionary argument was that the pursuit of commerce distracted people from spiritual matters, which were more important. In 1836, the Reverend N. Murray of Elizabethtown, New Jersey, reviewed a book on the "decline of religion in Christian churches." In the course of his review, Murray lamented the growth of a "worldly spirit" inside the church in the United States. He noted that Americans were "a people of boundless enterprise," with "canals and railroads ... cutting up our vallies, and running along our rivers, and penetrating our forests, and scaling our mountains, and stretching over our prairies." Such transportation improvements had a profound effect on Americans: "The country village whose very houses sleep to-day, is all excitement to-morrow." It was impossible to get distance from the pursuit of wealth. This led to a "public mania for the acquisition of wealth," which had even reached into the church itself. Across the country, there were members of the church and ministers who were "out in this giddy pursuit of the world." By collapsing distance, railroads and canals made economic achievement possible everywhere, which led to a "thousandfold" increase in the "business and bustle of the country."[46] Murray saw this as profoundly disturbing. There was the danger, therefore, that in the rapid spread of economic opportunity, moral life would be lost.

Steam transit also possessed the power to end life, and this, too, was cause for concern in religious contexts. In 1852, an article in *Harper's* expressed concern about the cavalier attitude that Americans could have toward lives lost in accidents. Titled "Victims of Progress," the article noted that whenever there was word of an accident on a steamboat or railroad, the inevitable thought was that the victims were "martyrs of an ever-advancing, never-finished civilization,—they die that steamboats may be better built, that railroads may be better laid, that the speed of traveling, by land and sea, may be accelerated in a ratio which never becomes constant, and toward a maximum which is never to be attained." The author averred that while some deaths do lead to the advancement of knowledge, there was the danger that the desire to advance would never be satisfied, and thus the deaths would continue: "It is all transition—movement evermore. Steam brings us no nearer the consummation than oars and sails. Newspapers, and railroads, and magnetic telegraphs hold out no better prospect of a resting-place, than the discovery of the alphabet, or the first invention of the art of printing." The drive for more was insatiable, the author warned.

The author did not want to be seen as opposing progress and allowed that it was "a most pleasant and desirable thing to be carried smoothly and safely 150 miles in four hours." And yet the author worried about the moral calculus that might lead to disregard for the number of people killed by steam. If "an all-pervading secularity becomes the predominant characteristic of our civilization—if science usurps the homage which is only due to religion," then "may it indeed become a grave question whether such a physical advance is, on the whole, a true progress of our humanity,—a progress tending *upward*, instead of horizontally and interminably *onward*."[47] The pursuit of technological superiority would lead to an indifference toward the lives of others. For this writer, it was critical that Americans tend to their spiritual needs and not value progress for progress' sake above all else.

Steam transit corporations realized the power that ministers could have in their local communities. Therefore, they tried to recruit ministers to their side, to head off the types of concerns outlined above. In 1838 the Western Railroad distributed a broadside to ministers in its effort to secure funding from the state of Massachusetts. The broadside noted that in order for the railroad to obtain this funding, the directors felt that the "whole People of our beloved Commonwealth" needed to hear how wonderful railroads truly were. The broadside encouraged ministers to spend some time in their sermons commenting positively on the "Moral effects" of railroads.[48] Thirty-three ministers published a response in the *Boston Recorder*, declaring themselves to be "profoundly sensible of the influence" that railroads would have "not only on the commercial concerns of this country, but also upon its social, moral and religious condition." The ministers cautioned, however, that they hoped that the construction of the Western Railroad would not have "untoward influence" on the citizens of Massachusetts. Specifically, they desired that when the railroad would be completed, it "shall in no degree interfere with the strict observance of the Christian Sabbath." Noting that the Boston and Worcester Railroad already carried mail and passengers on the Sabbath, the ministers expressed their concern that a similar result on the Western Railroad would be "highly disastrous to the interest of public morality, and to the eternal welfare of multitudes of their fellow citizens."

The ministers concluded that they hoped the board of the railroad would consider this proposition seriously, since they had shown such interest in public morals by asking the ministers to speak on the subject.[49] While perhaps not the response that the corporation was hoping for, it did demonstrate that religious leaders were taking the corporation's request seriously. The corporation hoped for the approbation of the community and knew that religious leaders were in a good position to convince the public of the righteousness of their enterprise, and religious leaders used the opportunity to express one of their leading concerns, about the sanctity of the Sabbath.

Pastor L. F. Dimmick of Newburyport, Massachusetts, recalled this request from the Western Railroad when he spoke in 1841 on "The Moral Influence of Railroads." Dimmick found much to praise with improved transit but ultimately struck a middle view. For Dimmick, the crux of the question of morality was this: "If this great method of locomotion . . . shall be so managed as to manifest due reverence for the God of the universe, and the institutions he has ordained, then its MORAL INFLUENCE, it needs no prophet to foretel, will be good—eminently good." Interestingly, Dimmick did not make a judgment on the railroad itself but rather how it would be "so managed": if the people running the railroad did so in a manner that was in keeping with Christian morality, then there was a chance for the railroad to have a positive moral influence. But, conversely, if "this distinguished improvement of the age shall be used in such a manner as to do dishonor to God, . . . then its moral influence, as every one can easily see, must be evil—eminently evil." He agreed that railroads constituted a great achievement, but he urged the directors of railroads not to forget that it was God who enabled their railroads: "His are the powers of nature, unlocked in the STEAM. His is the SCIENCE which planned, and the SKILL which constructed, your great work."[50] The railroad was evidence of God's incredible powers, and railroad companies should respect that through their operations.

Another minister who received the request from the Western Railroad, Rev. Dr. Gannett, gave a sermon over a decade later, in which he recalled this particular request and the response to it. He noted that only two ministers gave sermons at the time of the original request. Reflecting on railroads and their expansion in the subsequent years, Gannett ended up at a similar place to Dimmick. On the one hand, he hoped that railroads would improve relations among peoples: "It is not easy to exaggerate the importance of the continued intercourse that will now take the place of infrequent journeys, in lessening the probability of any hostile display." Railroads also brought benefits to individuals: "they enable the inhabitants of a crowded city to breathe the fresh air and behold the broad horizon of the country, to revel in the enjoyment and receive the instruction of nature." This was specifically an opportunity for people to feel close to the natural world as created by God: "Every one must admit that both bodily and mental health are promoted by ... communion, however brief, with the free and fresh influences of God's creation."

Gannett cautioned, however, that railroads could also have a deleterious influence. The constant moving about was leading to a loss of stability in life: "We are losing our attachment to home. We are less domestic in our ways every year.... There was too much excitement in our life ten years ago—what must it be now?" Steam transit also brought with it considerable wealth, and that wealth could itself have a positive or negative effect: "It may nourish selfishness, worldliness, and irreligion, or it may render its possessors more humane, more righteous, and more godly." Railroads, then, were not inherently good or ill but could facilitate either positive or negative results. Christians had to be constantly on their guard to ensure that technology remained a tool for good rather than the road to sin.[51]

For many writers, the objection to transit was often not an objection to technology per se but concern over what an obsession with technology reflected about a society's larger priorities. In 1835, the *Religious Intelligencer* complained that people were more willing to spend money on building things in the world, which are "only temporary," rather than to spend money on mission work to spread the Gospel. "Railroads are built—inclined planes are thrown over the towering mountains—canals excavated through solid rock—bridges thrown over rivers," the magazine groused, "and yet, for the salvation of a world, not enough is given annually to build and send out five ships."[52] For this writer, the country's priorities were incorrect. In 1839, a Dr. Riddle gave an address to the young men of Pittsburgh. He asked, "What have we gained by our rapid locomotion—our lightning speed of transportation—our ramifying rail roads—our meandering canals, if we lose the Sabbath? It is the soul of a nation's morality and religion."[53]

Orestes Brownson made a similar argument in 1844 when he railed against the impulses of society that valued commerce over the heart: "When we speak of the progress of the age, attempt to prove our advance on past ages, what are the facts we cite? Is it our progress in humility, in purity of heart, in a devout, reverential spirit? No; we point to our institutions for the multiplication of national wealth, to our industrial achievements, our steam engines, spinning jennies, steamboats, railroads, locomotives." Brownson felt that this calculus was backward, and that a culprit of this maldistribution of value was steam transit.[54] And J. G. Adams gave a lecture to the Bleeker Street Universalist Church in 1847 in which he allowed that railroads might "band together distant regions," but these and other improvements "are all lighter than vanity when compared with the moral education and progress of our race."[55] True value lay in morality, not the pursuit of worldly things. Some religious writers, then, warned against valuing steam transit above all else.

Far and away the most sustained complaint that religious thinkers had about steam transit was its capacity to despoil the Sabbath. Staunch preservation of the Sabbath had a long history in the United States, with each colony having a law against breaking the Sabbath.[56] As the country grew, religious reformers in the early nineteenth century paid special attention to the preservation of the Sabbath. This came up directly against the desire of many transportation companies to operate as often as possible, regardless of the Lord's designs on one day of the week. Generally speaking, even the ministers who were the most fervent advocates of railroads and steamboats urged them to take a day of rest one day per week. There was the occasional exception if the purpose was linked to religion: for example, the *Church Advocate* praised the Cumberland Valley Railroad in 1852 for allowing a reduced fare on the Sabbath to allow people to attend a religious revival camp meeting.[57] But generally religious writers balanced their enthusiasm for steam transit with a desire to see it stopped on Sundays.

The religious objection to traveling on Sunday was deadly serious. "Should you, by the urgency of business," warned one writer in 1835, "be tempted to ride in the coach, or sail in the steam boat on the holy Sabbath, remember if you should yield, you have aimed a blow at one of the mightiest pillars of your country's liberties; you have invoked the judgment of Heaven, and you cannot expect but desolation in your own soul."[58] Writers counseled Christians to not even associate themselves financially with violators of the Sabbath. Rather, they should use the tools of the businessman to promote Christian ends. "The stock of our railroads, and in our steamboat lines is much of it in Christian hands," counseled the *Essex North Register* in 1836. Christian stockholders had a special obligation to ensure that the Sabbath would not be abjured. "Amid winds and waves, and the hissing of steam, and the rumbling of cars, and smoke, and clouds of dust, and din of business, you must stop," the paper commanded. "You must withdraw your capital from establishments which violate the Sabbath."[59] By invoking the aural and visual upheaval of steam transit, the newspaper emphasized how disruptive transit could be to a contemplative Sabbath.

Some passengers were also concerned about Sunday travel, even as they traveled on Sunday. In 1849, J. Obear wrote that he would have liked to have stopped and visited a town while on travel, but "as we arrived there this Saturday morning I should have had to remain two days, or travel on Sunday which is against my principles." Therefore, Obear elected to "push on" and get closer to his final destination without having to travel on a Sunday.[60] Others were open to traveling on the Sabbath, even if they expressed hesitation. In 1852, Esther Belle Hanna wrote of her journey west from Pittsburgh, including a notation that "yesterday was Sabbath, but there is no Sabbath on the river at least one would think not. Our Captain is an unprincipled creature, however he gave permission for Mr. Hanna to preach in the morning. The sermon was listened to with attention & apparently much interest."[61] Sabbath travel was troublesome, therefore, but sometimes unavoidable. The possibility of introducing a church service on board a steamboat helped alleviate the mental anguish of breaking the Sabbath.

Some hoped that the increased speed of travel, made possible by steam, would help protect the Sabbath. The *Sheet Anchor*, published in 1844, believed that "railroad speed is rendering Sabbath travelling needless and unprofitable."[62] In 1852, the *Alabama Planter* commented that "railroads are as much His work as the Sabbath, and the two things will not be found to annihilate each other.—The Sabbath is now a more imposing and better observed institution than it ever has been."[63] But these hopes ultimately proved overoptimistic. While some corporations limited their work on Sundays, by and large travel on the Sabbath continued, prompted by the necessity of carrying the US mail. This was profoundly troubling to Sabbatarians. In 1841, a book reviewer reviewing a book about the Sabbath noted, "Canals and railroads, as well as our rivers, are becoming highways of iniquity" with their violations of the Sabbath.[64] In 1844, an Ohio Sabbatarian convention concluded that "commerce on the sea-board, and on our rivers, and on our canals and turnpikes, and railroads, is putting in motion a secular enterprise which is fast deadening the national conscience, and rolling the wave of oblivion over the sacred day."[65] The *National Era* told the story of a preacher who was obligated to travel on a Lake Erie steamer on the Sabbath. When he requested the ship's captain if he could hold a religious meeting, the captain replied, "No—for any minister who would travel on Sunday is not fit to preach on board my boat." The newspaper pointed out the hypocrisy with the captain's stance: "A minister of the Gospel might, as the Saviour did, find it necessary, and therefore innocent, to travel on a Sabbath day; but a steamboat captain, if competent for his place, could easily earn a living without habitually running a boat on that day." Thus, what the captain hoped was a witty retort was turned against him.[66]

Sabbatarians, of course, were not ultimately victorious in the antebellum era; they would not achieve their goal of ending Sunday mail until 1912.[67] They were unable to convince their fellow Americans to give up the convenience of Sunday travel. And they had critics. The anonymous author of *Prohibition of Sunday Travelling on the Pennsylvania Rail Road* asked why, simply because some people opposed Sunday travel, should they have "a right to insist that others who have no such convictions" obey the same rules.[68] Cartoonists also picked up on this theme. In 1855, an anti-Sabbatarian cartoon poked fun at things that were not allowed on Sundays, according to Sabbatarians. In the background, a steamship sat idle next to a sign that read, "This river is stopped from running on Sunday."[69] In another cartoon called "Sunday Laws," from 1859, a cat is about to be hanged for killing a mouse on Sunday. The preacher intoned that the cat had "eaten, drank, and slept, on the sabbath day; . . . disturbed my congregation . . . on the sabbath-day; thou hast tortured this poor little innocent [mouse] even unto death; all on the sabbath-day; and for these manifold transgressions thou deservest death." Nearby, a man with a horn lifted up to his ear complains, "I can't hear a word the minister says on account of them cars."[70] This cartoon exposed the hope of Sabbatarians versus the reality of Sabbath observance. Sabbatarians could protest against whomever they wanted—even a cat—but the rest of the world went about its business. Sabbatarians were vocal about their demands and remained so throughout the antebellum era. But they did not achieve national victory during this time, nor were all Americans convinced of their program.

Thus, there was ultimately a tension for antebellum religious thinkers regarding steam transit: technologies could be both positive and negative, depending on how they were used. The cautionary tone struck by the *Youth's Penny Gazette* in 1850 demonstrates this balancing act. The newspaper wrote, "The *railroad* is a blessing or a curse then, according to the USE to which it is put. Let us remember this."[71] Religious thinkers saw the technology as holding the possibility for extending the work of religion on earth but also the possibility of leading moral men and women astray. Christians had to be ever vigilant to ensure that they did not veer onto the wrong path.

Comfort after Accidents

Perhaps nowhere in the context of steam transit was the role of religion more profound than in helping Americans to wrestle with the implications of transportation accidents and the death that resulted. People groped for meaning when accidents happened. Most trips were safe, but some ended in horrific conflagration. In an accident, one person might be killed while his or her seatmate would be spared. The violence done was seemingly at random. Such events could leave families and communities reeling. As people gathered at funerals to mourn their losses, preachers helped their audience make meaning out of the accidents that had struck down their loved ones. In so doing, preachers often commented on steam technology and its role in society. There was, of course, death at the hand of transit prior to the development of steam transit. In shipwrecks, ministers followed a similar vein that we will see here: shipwrecks were seen as "retribution for sin, a chance for redemption, or a reminder of God's power."[72] But steam transit, as a human invention, added the risk of idolatry: of people "transfer[ing] their reverence from God to an object of their own making."[73] For these preachers, accidents and the death that resulted were signs that God's power overwhelmed anything that man could do, and a reminder that one should not put too much stock in the creations of man.

In 1837, Thomas Smyth preached after the destruction of the steamboat *Home*. Traveling from New York City to Charleston, South Carolina, the ship had wrecked off the coast of North Carolina on October 9 that year, killing scores of passengers. Steamboat travel was clearly familiar to those in Smyth's congregation, because Smyth felt that it would be easy for his audience to visualize the people on board the boat: "We can see this multitude of fellow beings, as they crowded on board that packet which was to restore them to their own sweet homes. We can accompany them as they cheerfully endured all the trials of their way, in the glad promise of a speedy voyage." Smyth knew that the congregation would be troubled by the destruction wrought by the accident and wonder how God could allow such an event to take place. He addressed that directly and drew out the lessons that were evident to him. The first was that God was severe as well as loving. Smyth noted that while people readily believe in a "God all merciful and gracious," his audience must remember that God is also "powerful as well as merciful." Thus, the accident demonstrated "the severity, as well as the goodness of God; on them who perished, severity, but towards those who escape, goodness, if they will be led by this goodness to repentance, otherwise they also shall be cut off." God's goodness gave the survivors another opportunity to live a better life.

Smyth's first lesson, that God was powerful and severe as well as good, led directly to the second: that those spared by God's goodness must take the lesson of God's destructive power to heart. Steam transit represented a remarkable technological accomplishment, but ultimately God demonstrated that He was still in control of the world: "Be taught, O man, thy impotence. Realize thy helpless dependence upon the omnipotence of Jehovah." The final lesson Smyth drew was that death could come at any time, and thus people should always be prepared to die. "Behold the fashion of this vain and transitory world passeth away," Smyth warned his parishioners. One should attend to "the absolute necessity of being now and always ready and prepared to die."[74] Smyth did not object to internal improvements or dissuade his flock from using them. Indeed, he opened his sermon with a visualization of travel and counted on his audience being able to easily imagine the scene. But the destruction of the steamboat illustrated that no matter how man's achievements might dazzle, ultimately power remained solely with God. Accidents on a steamboat could strike seemingly at random and without warning, catching saint and sinner alike. Thus it behooved the Christian to prepare his or her soul and always be ready for the moment when death would strike.[75]

The destruction of the steamboat *Lexington* in January 1840 summoned forth another series of religious reflections. The *Lexington* was owned by the famed industrialist Cornelius Vanderbilt, and when it sank all but four passengers perished, with nearly 140 dying in the accident. A publication on the public inquest following the accident noted that it was a "calamity scarcely paralleled even in the mournful catalogue of steamboat disasters." One writer addressed the painful question no doubt felt by those who lost loved ones: why was my husband or child killed, but others were spared? The author drew a straightforward lesson, that such questions were for God to know the answer to even if humans could not understand them. The only solution was to "bow in humble acquiescence to His will, who is too wise to err and too good to be unkind." Like Smyth, this writer drew lessons from the steamboat accident. "To those who are living for the world only," he noted, an accident "proclaims, as in thunder tones, 'Boast not thyself of to-morrow, for thou knowest not what a day may bring forth.'"[76]

Religious leaders were happy to see steam transit as a celebration of man's accomplishments, but accidents brought forth the lesson of humility. John Stone also preached a sermon on the burning of the *Lexington* and reached a similar conclusion. He noted that Christianity demanded that people "live in a more habitual readiness to be separated, at a moment's warning, from those we love; and it would have us love our *Saviour* better, and more fervently, than either father or mother, either wife or children." Stone concluded, "Christians, awake, and in holy earnest get ready for your departing. *Your* turn may come next."[77] Again, the message was clear: death could come at any moment, and Christians had to be ready for that moment to come. Another eulogy from 1852 reiterated the same theme, that the death of friends had a message for those who remained behind: "When the servants of Christ are thus early and suddenly called away, there is this clear voice to surviving relatives and friends, from out [of] the afflictive dispensation. It is a call to them anew, to set their affections supremely there, where yet another treasure has been gathered in of God, as it were to draw their hearts on after."[78] In this case, the minister tied death to a powerful message: those left among the living should work to set themselves right before God before their own death came.

These lessons could strike home with the audiences. In May 1853, Anna Marie Resseguie reported in her diary that her local preacher discussed a recent disaster on a railroad, which killed dozens of people. As usual, the minister's intent was to exhort his flock to prepare for the fact that death could come at any time and God's judgment was sure. Resseguie noted that the minister said, "We ought to think often of death in order to be prepared for it." The following year in October, her minister preached again on a catastrophe, referring to two vessels that ran into each other off the coast of Newfoundland: "Mr. Clark spoke so well of the dreadful calamity of which we heard this week, that it was God who had permitted it, the carelessness of the crew and everything. It might try our faith to the utmost. It was the only comfort, all that would give peace, to feel that it is a righteous God who overrides all events." A little while later she wrote again, on the fact that several preachers had taken accidents as their topic: "The ministers of N. York chose for their topic this calamity on Sunday. They all acknowledged God's hand in it. By no mere chance could two steamers start from opposite shores and come into collision on the wide ocean. . . . All agree that in this sad calamity God is saying to each one, 'Prepare to meet your God.'"[79] Accidents were to teach Christians to be prepared at all times for death, a lesson that resonated with listeners.

D. R. Brewer preached on the loss of the steamboat *San Francisco*, which suffered an accident in January 1854. He warned his audience that such accidents carried a message from God: "Does He not tell us, for one thing, that we have talked too much, in our times, about man's triumph over nature, the conquest of matter by mind, and all that we have done and are going to do, in that line?" Brewer charged, "Hardly a speech is made by any of our public orators, without a high flight in this direction. Our amazing advance beyond past generations, our numerous and increasing inventions, our railroads and telegraphs and steamers and clippers, our *fast* things,—of all sorts—these are unfailing topics for brilliant rhetoric and interesting conversation." But boasting of such things was "a denial of the Lord God," and Americans were in danger of worshipping "manufactured gods; gods of iron and wood; cotton-gins, spinning-jennys, locomotives and steamships."[80] The accident was a reminder that people should not worship these false idols. On January 15, 1854, Caroline Barrett White, in her journal, also described the loss of the steamer *San Francisco*, demonstrating that she had inculcated the message given by different religious leaders about the meaning of these accidents. She mourned someone who had perished on the ship but also noted that "probably he, whom we mourn, would not exchange the blissful glories of the 'better land' for a few more short years on earth."[81] The person had perished but was moving to a better place.

In 1855, there was a significant accident on the Camden and Amboy Railroad near Burlington, New Jersey. F. Reck Harbaugh gave a sermon in Burlington on the topic of the accident. Harbaugh knew that railroads held an appeal, and he alluded to their aesthetic magnetism: "There are few sights more engaging than the fearful and graceful motion of a train of cars—and the proof of it is this, we never tire of looking at them." Graceful, but also fearful given the accident they were there to mourn. Harbaugh believed that the very location of the accident sent a message that the people of Burlington must heed. Although people may have felt that the deaths of so many people was harsh, Harbaugh encouraged his audience instead to see God's *"mercy,"* since God caused the accident to occur in such a place that put "sufferers within reach of almost instantaneous relief." The message of this disaster was that it was time to "put away our unbelief and impenitence, and turn unto God in the deepest contrition." Under such circumstances, "sudden danger will be perfect *safety*. Sudden death, but 'going home.'"[82] The image of going home to God comforted those who heard it that these deaths—although sudden and painful—were not meaningless.

Another sermon after the Burlington disaster was given by Henry Boardman in Philadelphia. Like others, Boardman emphasized the lack of human knowledge and agency: "There is *very much* about a disaster like this, which mocks our wisdom."[83] The underlying message was the same: Christians should prepare their souls for death at any moment, which these accidents demonstrated was ever a possibility. Accidents such as the one in Burlington presented ministers with the opportunity to drive home a message about preparation for death but also to provide a message of comfort.

In 1856, Lemuel Porter gave a sermon at the funeral of Charles W. Nichols, who was an engineer on the Western Railroad. Porter gave a warning to his listeners: Nichols's "poor lacerated body" was a "sad picture of what any one of you may be within the next twenty-four hours."[84] Again, steam transit accidents demonstrated that life could be snuffed out in an instant, and the grisly visualization that Porter provided was a stark lesson for his congregation. In 1857, J. B. Shaw gave the eulogy for an engineer who had been killed in a train that was apparently sabotaged in some way. Because of this, the pastor had strong words for whoever caused the engineer's death, saying that the public should give "undivided indignation" to the person who caused the damage. Nevertheless, he continued the theme of other pastors: the death of the engineer should remind all who work for the railroad to be in "constant preparation." He warned his audience, "None can tell who may be the next victim, none tell whose wife may be the next widow. Be ye also ready—for in such an hour as ye think not, the Son of Man cometh! Verily, there is but a step between you and death. So live that it may be but a step between you and glory."[85]

For Porter, Shaw, and their colleagues, delivering such eulogies for people who had been denied a Good Death was an opportunity to warn everyone else in their congregation. Death on steam transit could be sudden, brutal, and random, and the victims could be isolated from family and denied the opportunity to deliver the last words to those they loved. With so many of the aspects of the Good Death missing, good Christians were urged to live their lives as to prepare for death at every moment. Travelers accepted those risks. But the challenge of the lack of the Good Death wrought by antebellum steam transit served as a preview for what horrors would come in the Civil War.[86] The cultural groundwork for dealing with mass, unexpected death on the battlefields of Gettysburg and Chickamauga was laid in the antebellum era on railroad tracks and steamboat docks.

Steam transit touched many aspects of American life, and religion was no exception. While there was some tension among Protestant writers and preachers, these religious thinkers largely embraced the technology. Steam transit presented real benefits, was a sign of God's favor, demonstrated the superiority of Protestantism over Catholicism, and could be used to spread the Gospel. There were, of course, some reasons to be wary. Technology could encourage the pursuit of wealth and Sabbath-breaking, which could distract people from Christian values. But the overwhelming Protestant response was to see steam transit as a means to spread the Gospel, not a hindrance to living by the Gospel. As accidents happened on water and rail, people turned to their ministers for answers as to how seemingly random events could strike down their loved ones at any moment. In technological accidents, preachers saw a powerful lesson that God remained in control of people's lives at all times and that death could come for any person at any moment. Thus, all had to be continuously prepared for death by living as good a life as possible, since death could strike at any moment. As steam transit expanded across the United States, therefore, religious leaders moved to both understand it and incorporate its lessons into their broader teachings. The move for understanding led them to embrace the possibilities that steam transit offered for disseminating the Gospel and to comfort their congregations when accidents created painful losses. Examining religious leaders' response to steam transit demonstrates some of the tensions of the new technology. Steam presented real benefits: it could help spread the Gospel and was a sign of God's favor toward the United States. But there were also challenges: steam transit led people to violate the Sabbath and created a need for eulogies as accidents claimed lives and Americans searched for answers about a changing relationship to death. The degree to which steam transit was discussed in a religious context highlights the degree to which steam transit spread through American culture.

CHAPTER FIVE

Black Passengers

No group of people excited more controversy in the antebellum era with respect to travel than Black Americans. Whether in the North or South, or concerning enslaved or free people, the issue of Black mobility elicited wide reactions from the white population. We have seen how for white passengers, the growth of steam travel led to transportation becoming a commodity. When Black passengers were concerned, however, these issues must be reassessed. Not all Black passengers could experience commodified travel. In the South and North, norms about who could sit in what part of a train or steamboat were enforced by conductors and captains who wielded a great deal of power during a journey. The examples in this chapter illustrate the degree to which white Americans worked to restrict Black mobility, and how Black people pushed against these strictures and used steam transit to their own advantage.

Black Mobility

The question of Black mobility in the antebellum era was profoundly linked to slavery. Antebellum Americans were acutely aware of this point. For example, in his autobiography, Frederick Douglass noted that slavery forced enslaved people to be a "fixture." The decision to move enslaved people from one place to another rested in the hands of their white master. An enslaved person could have "no choice, no goal, but was pegged down to one single spot, and must take root there or nowhere." Of course, a master's decision could break up a family; thus "the idea of removal elsewhere came generally in the shape of a threat, and in punishment for crime. It was therefore attended with fear and dread," Douglass concluded. Paradoxically, for enslaved people, movement did not represent freedom; rather, it meant that they could be ripped from their loved ones. Douglass contrasted this feeling of threat with the feeling of liberty that movement gave to free people. A free person could, for example, strike out to the West and rebuild their life or start anew. Movement meant possibility. But for the enslaved person, movement did not mean possibility. Their mobility was limited by their master's plans for them, and worse, mobility could be a form of punishment.[1]

Douglass wanted to draw a stark image for his white audience, many of whom presumably never gave a second thought to their ability to travel. As we have seen, for white people travel meant seeing friends and family or conducting business. While travel had a different meaning for enslaved people, the introduction of steam transit to the southern United States did alter the possibilities for enslaved people. Douglass was correct that enslaved people had their mobility constrained by their masters. Yet at the same time, the same masters who wished to constrain Black mobility needed those same enslaved people to travel in order to further the business of the southern economy. As historian Susan Eva O'Donovan has noted, "travel on a master's command was part and parcel of a day's labor" for many enslaved men and women. Arguments about Black mobility filled legislative debates, legal cases, and the columns of newspapers. Enslaved people were both constrained and mobilized by their master's will: whites had no desire for enslaved people to escape but did need them to move to drive the economy forward. Inevitably, the opportunities to move about—even in a constrained local environment—opened up possibilities for the enslaved person to make an escape.[2]

Thus, while Douglass emphasized fixture for his antebellum audience, he was himself evidence of the possibility for escape that steam aided. For the enslaved people who ventured off of the plantation, railroads and steamboats suddenly offered a much quicker route away from their master's land. Steam transit must have offered an enticing prospect for anyone contemplating escape. Douglass used steam transit in his escape from slavery, and many other enslaved people did the same. Railroads and steamboats threw into relief another aspect of slavery: that the maintenance of white power was dependent in part on local knowledge. If an enslaved person was a "fixture" in a community, he or she was thereby known to the local white populace. Therefore, if an enslaved person attempted to escape on foot, they would be easily spotted and recognized by their master's neighbor. But if an enslaved person could get far enough away from their master's property, they could more easily convince people who did not know them that they were legitimately traveling, or perhaps even move without being recognized. Steam travel helped expose how dependent slavery, as a system, was on local knowledge for enforcement. Removed from local context, enslaved people had a better chance of fleeing their enslavement. Southern state governments recognized this immediately and worked to govern the movement of enslaved people on steam transit and punish the white people who assisted them. Southerners also fought for federal legislation to enforce northern assistance in maintaining slavery.

Enslaved people quickly realized that steam transit had the possibility to take them far from slavery—and it could do so rapidly. The risks were high (needing to hide in confined quarters or hoping to hide in plain sight with disguises and accomplices), but the potential payoff (swift passage out of the South) was tempting. In the history of technology, historians often consider how individual users of technology bend it to their own ends, unintended by an inventor or a creator of the technology.[3] Surely no southern railroad corporation or steamboat line ever *intended* that their technology would become the means for enslaved people to break free from bondage. But as we will see, either alone or with the aid of others, many enslaved people made the attempt. Enslaved men and women across the South exercised courage and ingenuity to bend steam travel to their own designs. While we often only know these stories because escape attempts were foiled and publicized, such attempts nevertheless point to the inventiveness of the enslaved people and their steadfast desire for freedom. Ironically, enslaved people played an integral role in the construction and operation of steam transit. Thousands and thousands of enslaved people toiled across the South to build and service the region's transportation infrastructure.[4] In so doing, they were literally building and operating the mechanism that many of their fellow enslaved people would use in their bid for freedom. As the *Massachusetts Quarterly Review* noted in 1849, "Railroads which have sunk, in mercantile phrase, a great amount of money thus invested at the South, have also, in some few instances, rendered the living 'property' singularly insecure."[5] When the white South adopted steam transit to improve its economy, it created a mode for enslaved people to seek their freedom at the same time.

As enslaved people made their way north, points of transit became areas of public controversy. If a gang of white people pursuing an enslaved person caught that person at a railroad station or on a wharf, public confrontation could result. Crowds gathered and—depending on the place or its political leanings—could band together to fight for the captives' freedom. In that moment, for any people standing nearby, slavery was at once transformed from an abstract political argument to something quite real: this person, in front of me, could instantly be whisked back to slavery by the next train or steamship. Steam transit did not just carry people and goods but also crystallized political controversies in the towns along its routes.

For the free Black people of the North, the stakes were different. While free Black people did not live in slavery, they nevertheless faced widespread discrimination. Sometimes this was explicit in the regulations established by a corporation. In other cases it was more capricious, at the hands of a cruel conductor or steamboat captain, or in response to complaints from other passengers. Black writers would note how white people in a boat or railroad car would silently approve of ill treatment by looking away and refusing to speak up on behalf of those suffering discrimination. The maintenance of racial hierarchy required broad community participation. In this chapter, we see examples of white passengers both protesting the mistreatment of Black passengers on steam transit and, far more often, sitting silently while Black passengers were verbally abused or physically assaulted. In either case, the larger group's acquiescence or action helped determine how Black passengers were treated.

Even if enslaved people were not present on steam transit, the topic of slavery was never far from the lips of passengers. Based on descriptions from contemporaries, it seems that steamboats and railroads were abuzz with debates about slavery. Just as religious leaders praised steam for helping them spread the Gospel, steam travel was also a useful tool for abolitionists, as it allowed them to more easily promote their cause. One abolitionist urged another to take New York City as a base in 1837, thanks to its convenient steam connections: "Half the moral power of the nation lies with[in] 24 hours easy ride (mostly steam boat) of New York City."[6] Abolitionists seized on steam transit as an opportunity to spread the abolitionist message in unprecedented ways. Chance encounters with slaveowners on trains and impromptu lectures on steamboats all provided opportunities to share the abolitionist message.

The first two decades of the twenty-first century saw a significant growth in the scholarship on Black mobility and the evolving responses to fugitives in different parts of the North.[7] This literature has created a more nuanced understanding of the actual operation of the Underground Railroad. Despite its name, the routes for escape were dependent on aboveground steam transit, in both North and South. Historians such as R. J. M. Blackett and Robert Churchill have carefully mapped the geography of the North to assess local attitudes toward fugitives and the ease of moving through the North. That ease was aided by steam, and the expansion of steam transit (both the railroads implied in *Underground Railroad* and steamboats) changed the ways in which enslaved people escaped to Canada. Thus, the constant expansion of steam transit in the United States prior to the Civil War helped accelerate the urgency of the discussion about the future of slavery. Steam transit made it easier for enslaved people to flee, and newspapers were full of stories about escapes or attempts. The expansion of steam transit brought the issue of fugitive slaves to the fore by easing the process of escape for many.

In my previous book, *Railroads in the Old South*, I highlighted the fact that white southerners wanted to both retain their racial hierarchy and adopt the most modern form of transportation available.[8] Taking a closer look at the experiences of people who used steam transit as a means of escape complicates the story more than I previously allowed, because I did not sufficiently analyze how Black people themselves were using these networks. White southerners wanted a modern transportation network, but in forcing enslaved people to build and sustain this network, white people ended up accelerating escape routes. North of the Ohio River, transportation networks were then major points of contention with respect to the return of those who had escaped slavery.

Not all escapes were successful, but there is no doubt that the expanding networks in both North and South hastened and intensified debates about fugitive slaves and the propriety of slavery. Moreover, although white and Black people used the same networks, their use was quite different. For white passengers, steam transit commodified travel. White passengers no longer needed to piece together their journey but could buy a ticket and let the company take care of the rest. For enslaved people attempting to use the steam network as a means of escape, however, travel retained its precommodified character. To be sure, there were some who attempted to purchase tickets and blend in with the crowd. But the vast majority of those attempting escape had to create their own routes, jump on and off trains before they arrived at stations in order to avoid detection, stow away on steamboats, bribe employees or hope to depend on their kindness, and use other techniques. Travel for those escaping slavery was not simply a matter of boarding transportation and riding to a destination; it required the ingenuity, planning, uncertainty, and hardship that characterized precommodified travel for white passengers.

Much of the material in this chapter is drawn from the Black press. Black newspapers reported routinely on attempted and successful escapes from slavery, as well as poor treatment received by Black people on transit in the North. Black newspapers also printed general interest items about transit, such as accidents, changing fares, or newly developed steamboat routes.[9] These newspapers not only were leaders in reporting on Black mobility but were part of the general interest in the antebellum era on steam transit. They also took a special responsibility to shine a light on the injustices wrought by northern companies.

Regulated Mobility

Steam transit played an integral role in the internal slave trade of the South.[10] Numerous contemporary observers commented on the gangs of enslaved people they saw being transported from one region to another. Steam transit facilitated the internal slave trade and thus the destruction of familial bonds. Douglass argued that the enslaved person's fundamental character was his or her rootedness, but for the southern economy as a whole, Anthony Kaye has argued that "mobility" was critical for slavery's "economic dynamism."[11] The ability to move enslaved people to new areas of the country was crucial for slavery's growth, and traders took advantage of steam transit networks to ply their trade. Slave trader A. J. McElveen informed a correspondent that the railroad was a "safe" way to transfer his enslaved people from one part of the South to another. Indeed, once he learned that the railroad had "Broken down," he wrote that he was "fearful It will be trobled to Get them down Safe." Later, he requested that an enslaved boy be sent by railroad and McElveen would "meet him at the R.R. depot in Sumterville."[12] Another slave trader in Richmond, Virginia, had an "enslaved assistant" who met enslaved people at the city's "various railroad, steamboat and canal landings" in order to transfer them to the trader's "mart, where they could be held securely prior to their sale."[13] For these traders, transferring enslaved people by steam was a reliable and dependable way to do business.

Railroad companies handled enslaved passengers in different ways. There was no single rule for the entire South: different companies had different regulations or accommodations, and different states had different laws. Enslaved people who were accompanying white passengers—for example, an enslaved woman with charge of a white child—might ride in the same car with the white passengers. Allowing such racial mixing did not sit well with all white passengers. In 1856, Edward Spann Hammond traveled on a railroad in Tennessee and recorded in his diary that he "was surprised at the conduct allowed mulattoes on the cars."[14] Some enslaved people also traveled independently on steam transit while performing work for their master. Historian Charles Bolton notes that slaveholders had to give such enslaved people passes, "which created an opportunity for fugitives to carry false permission slips."[15]

Larger groups of enslaved people being transferred on steam transit would be mostly shielded from public view if they were put in cars with much worse accommodations, as if they were freight. The South Carolina Railroad, for example, carried certain groups of people at lower rates, when it announced in 1850 that "Gangs of Negroes and Emigrants" would be "carried at reduced rates" if they took a specific train each day. Otherwise, they would be "charged full rates."[16] Lumped here with emigrants, another group handled by railroads as a lower class of passenger, enslaved people were primarily moved on railroads as "gangs," not individuals. Some companies saw the business of transporting enslaved people as valuable; historian Wilma Dunaway notes, "To attract slave trader business, the East Tennessee and Virginia Railroad implemented a policy to carry small slave children free of charge."[17] The company hoped that such an incentive would increase this specific type of business.

Although enslaved people were segregated into different cars when transported en masse, there were still plenty of opportunities for them to be observed when traveling.[18] In 1841, future president Abraham Lincoln witnessed a group of enslaved people on a steamboat. Their owner had purchased them in Kentucky, and Lincoln observed that they were chained together, "precisely like so many fish upon a trot-line."[19] In 1857, Henry Ashworth recorded in his diary that he saw fifty enslaved people board a train that he was on. Half of the people were men, who were "fastened together by hand cuffs," and the balance were women and children. The conductor told him that "they had already taken 6000 on at the rate of 1000 per week during the six weeks of the year which had then elapsed."[20] While this is an anecdotal response from one railroad, it does suggest that traders made ample use of the new technology when it was available.

The Black press also reported on groups of enslaved people transported by steam. The accounts noted the poor and uncomfortable accommodation that these groups received. In 1848, a writer saw a "drove of more than fifty men and women . . . closely packed into one car."[21] That same year, a writer reported from South Carolina having seen "fifty or sixty of these wretched creatures . . . huddled into an old box-car, without seats or any accommodations whatever, and fastened in, so that none might escape at the stopping-places, or throw themselves out of the cars and destroy their lives, in a fit of desperation."[22] In 1852, the *National Era* published an account of enslaved people being transported for sale by a steamboat that was so cold that a member of the boat's crew "threatened to raise a company to liberate" the enslaved people "unless they were placed in more comfortable quarters." Apparently the captain relented, and he "loosed" the chain holding the group together, "which allowed them to gather round the stove."[23] Years later, that same newspaper noted that "heavy shipments of Negroes for the far South are made almost every day by the Seaboard and Roanoke Railroad."[24] Historian Robert Gudmestad notes that some steamboats were even "modifi[ed] to accommodate the interstate slave trade," such as a boat that installed "two parallel chains that ran the length of the main deck and were bolted down." These chains, in turn, had "twenty-five short handcuffs extend[ing] from each chain."[25] Each of these stories demonstrates that the white South eagerly used the most modern transportation available in order to stabilize its racial regime.[26]

Reports of enslaved people transported by steam continued throughout the antebellum era, demonstrating that it remained an important way to move large groups of people.[27] In 1860, *Douglass' Monthly* published an account from someone in Alabama who saw "two car loads of negroes" attached to the train he was on. When he went back to the area where the Black passengers were being held, he found "some 150 negroes, and old men, women and children, mothers of large families, some alone, some surrounded by their offspring." These enslaved people were being taken from Virginia and North Carolina to the slave market in New Orleans and were transported in appalling conditions. The writer spoke to the conductor, who "frankly admitted that negroes, whom he could not mistake to be slaves directly from Africa, did frequently come on their road; that 200 such came the week previous, and that 800 more were contracted for."[28] The conductor's implication was that these people were being brought to American shores in violation of the ban on the international slave trade. Steam transit in America then facilitated this trade by speeding the enslaved people to their destination. Furthermore, as slavery moved west, enslaved people were transported by train and steamboat to slavery's farthest reaches.

In 1857, the Black newspaper *National Era* lampooned the practice of treating Black people as freight when it published a comical piece in which a Black woman was unable to produce the requisite fare. When she offered to pay the freight rate of five cents per foot—and thus pay ten cents for her two feet—the conductor replied that the passenger rate was five dollars. The woman countered, "Yas, yas, I knows dat, for white folks—folks *what am folks*—but I'se nobody; I'se *freight*, I is."[29] In this fictional story, the woman attempted to use white America's racism to her own advantage. Since the country did not consider her to be a person ("I'se *freight*"), she reasoned that she was then under no obligation to pay the rate reserved for people ("folks *what am folks*"). The story was published just months after the U.S. Supreme Court's *Dred Scott* decision, which ruled that people of African descent could not be citizens of the United States. By taking *Dred Scott* to its logical conclusion, she exposed the fallacy of treating Black passengers as less than human.

Free Black people worked as employees on steam transit, particularly on steamboats. For southern white people, this was extremely problematic. Such laborers highlighted for white people the dangers of steam transit: if enslaved people interacted with free Black people, they might be tempted to pursue their own freedom. As we have already seen, the design of steam transit in the United States promoted a great deal of mixing, whether passengers wanted it or not. For white southerners, this possibility constituted a peril. Passengers traveling through the region definitely noticed free Black people working on steam transit. One traveler wrote in his 1839 journal that he met a free Black man in a stagecoach outside St. Louis. The diarist described the man as formerly enslaved, who "acted as steward about Steam Boats" and lived in Indiana, where he had acquired some property.[30] Free Black people working on steamboats as cooks or freight handlers attracted a great deal of concern from slaveowners and southern governments.[31] Their mobility and ability to mingle with the enslaved population on wharves, streets, boats, and elsewhere was troubling to white southerners intent on enforcing racial hierarchy.

Throughout the antebellum era, states moved to regulate contact between free Black people and enslaved people. Several states "passed statutes requiring northern sailors of color to either remain aboard ship or be confined on land while their vessels were in port."[32] For their own part, enslaved people who were plotting their escape knew that waterfront work would present opportunities to escape on a ship. In 1853, an enslaved man named Isaac Forman, who had been working as a steward on a steamboat based in Norfolk, Virginia, escaped to freedom in Philadelphia via a different steamboat.[33] Working aboard a steamboat had created the means and opportunity for this escape. Historian Michael Thompson has characterized this waterfront work as an "ideal halfway house on a runaway's road to freedom."[34] Such employment provided ready access to a key means of escape.

The concern about mixing free and enslaved people was strong along the western rivers and particularly strong in New Orleans.[35] In 1838, a New Orleans newspaper advocated that all free Black people be immediately expelled from the city. The reason was the bad influence that free Black people could have on enslaved people. Those working on steamboats came in for particular criticism: "Doubtless many of us have observed the freedom with which free colored stewards and cooks of the steamboats talk of assisting slaves to escape from New Orleans to the upper country, and of their willingness to help them." This article underlined the transgressive role that free Black people could play, as outsiders to the area and with ready access to a fast means out of the region.[36]

Three years later, another New Orleans newspaper argued that free Black people working on steamboats were "a dangerous body of incendiaries."[37] Separately, a letter written by an enslaved person who had escaped was referred to in the newspaper *Ohio Statesman*. The letter made clear that these fears were entirely warranted, since the writer confirmed that "blacks on the steamboats would point out" abolitionists or others who could assist enslaved people in escaping.[38] In 1841, the St. Louis *Gazette* also argued against the employment of free Black people on steamboats, since they would encourage enslaved people to run away.[39] An 1849 report in Memphis alleged that "free black sailors on the steamboat *E. W. Stevens* had been 'recruiting' slaves in the city for passage to free states."[40] All of these accounts hint at the fluidity that characterized the free and slave states bordered by the Ohio and Mississippi Rivers.[41] Under the right circumstances, an enslaved person could find a free Black person to help them, or perhaps an enslaved person could use the fact that free and enslaved people mixed on the docks to more easily slip out unnoticed.

The question of mixing free and enslaved people persisted throughout the antebellum era. In 1860, an enslaved person escaped to Canada thanks to assistance from "his master's brother," who "was employed on a steamboat, and had promised long ago to help him escape, when a good opportunity should occur." When the moment was right, "the gentleman opened the hatch and told him to jump in. He did so and stayed there until the boat arrived at a free town."[42] This type of assistance is exactly what southern states wanted to crack down on. In 1860, a steamboat porter was "sentenced to be hung ... for assisting a slave in his attempt to leave the State on said steamer."[43] The penalty of death illustrates how seriously southern states took the threat of Black mobility.

Demands for laws to restrict Black mobility came from the local white community. In Charleston, South Carolina, a group of prominent residents formed a committee in 1835 to handle the "incendiary machinations now in progress against the peace and welfare of the Southern States." These machinations touched on steam transit. Their eleventh resolution demanded that the city's harbor master maintain "a correct list of all the persons arriving to and departing from this Port" and that the railroad company "have correct lists of all persons arriving and departing by that conveyance, whether white, free coloured, or slaves." The purpose of having these lists was so that they could be "regularly examined, to the intent that Incendiaries and other evil disposed persons coming amongst us, or attempting to pass through this State, may be detected and exposed." Enslaved people were not prohibited from traveling, but it was clear that some in the community wanted to keep all passengers closely monitored and expected private enterprises to play a role in enforcement.[44] A Virginia newspaper printed a similar complaint in 1836: "The Rail Road affords great advantages to runaways to make their escape, and our municipal authorities, should make such regulations as would secure the property of our citizens."[45]

A New Orleans newspaper made a comparable protest in 1855. Apparently, enslaved people were making a quick escape to the river via railroad and then boarding a steamboat to move to the North. The newspaper placed responsibility in the hands of private corporations, urging that "railroad agents and steamboat captains should carefully question any white man coming on board with one or two slaves."[46] Communities called for the government to hold private transportation companies accountable for when enslaved people used steam transit as a means of escape.

In response to such demands, legislation sprang up across the region. A Louisiana law from 1816 held captains of ships "liable for the value of slaves who escaped on their boats" and imposed a fine of $500. This law thereby protected the owners of the boat but made captains liable for enslaved people on their boats, however the enslaved people may have gotten there. Thus a law designed to impose restrictions on the movements of Black people in fact laid significant penalties on white people.[47] Other laws took a similar tactic. A Maryland law of 1839 "forbade any railroad, steamboat, or other vessel to transport an unaccompanied slave" and levied a $500 fine for violators.[48] In 1840, Louisiana set new penalties on steamboat captains if enslaved people were found on their vessels, and it was constructed in such a way that "the law creates a presumption" that any enslaved people on board without written consent "were received on board with the intention of depriving their master of them, and of transporting them out of the State." The boat captain here was guilty until proven innocent. The only way for a boat captain to overcome this presumption of guilt was to provide the testimony of two witnesses not employed by the steamboat.[49] Boat captains appear to have been able to defend themselves in court. Historian Charles Bolton found that the Louisiana Supreme Court "found for boat captains about as often as they awarded money to slave owners, influenced not only by the reputation of the captains but how they dealt with each situation."[50]

In 1840, North Carolina passed a law prohibiting the transportation of enslaved people "upon Rail Roads, Steam Boats, or Stage Coaches, without written permission from their owners."[51] In 1841, Virginia had a similar law that any "vessel or steamboat" transporting a fugitive "whether with or without" the knowledge of the boat's owner was "liable not only to imprisonment in the Penitentiary as a felon, but to the owner of such slave to double the value of the same," in addition to other fines. Thus, a Virginia newspaper cautioned readers, ship captains "cannot be too cautious in examining their vessels previous to their departure, to assure themselves that all is right."[52] For their own part, slaveowners advertising for the recapture of enslaved people routinely included in their advertisements an exhortation to ship captains not to "harbor or receive" an enslaved person "under the penalties of the law."[53] In the eyes of the law, the ship captains bore the burden for ensuring that their ship was free of enslaved people.

That same year, the state of South Carolina enacted a New York Ship Inspection Law, which "stipulated that vessels owned in any proportion whatsoever by a citizen of New York could not depart Charleston before undergoing an inspection for runaway slaves or fleeing criminals." Ships had to pay $10 per inspection as well as supply a $1,000 bond to cover any damages if an enslaved person was discovered on the ship. Although most slaveholders in the state supported the law as necessary to prevent their enslaved labor from escaping, some in the state opposed it, because it would drive up the cost of shipping from Charleston and thus potentially push business to other southern ports. Despite these complaints, the law remained in force for the remainder of the antebellum era.[54] Other laws worked to restrict contract between free and enslaved people. An 1830 law in North Carolina threatened thirty-nine lashes to any Black person who visited a docked ship that employed free Black people.[55]

Requiring private companies to take responsibility for any potential enslaved person on board was also the case in other states. In 1849, Georgia passed a law that made it an offense "for any conductor, fireman, engineer, or other officer or agent on or managing or conducting any Railroad in this State to allow any slave to travel on the same," unless there was a "written permit from the owner."[56] A Georgia court found in 1850 that if an enslaved person was on board a railroad "without the knowledge and consent of the owner, and he be injured by negligence or otherwise, the company will be liable, though the negro have a general pass."[57] Once again, the owner's consent was paramount, harking back to Douglass's speech that enslaved people could not control their own mobility. In 1856, the Alabama Supreme Court determined that "every railroad, steamboat or stage is liable to the owner of any slave absenting himself from his owner, who may be found traveling with them under any disguise or concealment."[58] The message across the South was increasingly clear. White owners feared steam transit as a means of escape, and they convinced legislatures to pass laws requiring private corporations to assist in the maintenance of their property rights.

In addition to legal strictures, some companies made their own moves to restrict Black people from traveling. In 1852, the Virginia and Tennessee Railroad published the following notice: "No Slave will be permitted to ride on the Railroad cars, without *special written* permission from his or her owner, or the agent or guardian of the owner, unless accompanied by the owner or person otherwise entitled to their service. Such permit must be given in duplicate, one of which will be retained by the Conductor of the Train."[59] Likewise, in 1856, the East Tennessee and Georgia Railroad published a rule that "Persons wishing their Negroes to travel on the Cars of this Company must procure Tickets for them in person, or through some reliable person well known to the Agents. Written permits to pass will not be regarded as sufficient authority, as they are often forgeries."[60] These corporations did not wish to be liable for any enslaved people who escaped, and thus the corporations required definitive proof of the people's enslavement. But this regulation also illustrated how slavery was dependent on local knowledge, and the mobility of enslaved people threatened that knowledge. While pro-slavery rhetoric considered slavery the appropriate state for Black people, this regulation revealed how tenuous slavery was. A person's status as an enslaved person was not immediately apprehensible or "obvious" from skin color: it had to be verified, either in person by someone "well known" to the corporation's employee or in writing and in duplicate.

All of this corporate and legal activity to restrict mobility and hold steamboat captains and railroad conductors accountable was driven by the fact that the enslaved people were using steam transit for their own ends: to escape. The mere presence of steam transit did not stoke white fears and mean that enslaved people could escape. Rather, enslaved people forced the issue by daring—and even succeeding—to escape. Laws, policies, and fines were all in place in hopes of constraining Black resourcefulness and creativity, and enslaved persons and their free Black compatriots pushed back against these constraints.

Mobilizing to Escape

Douglass identified rootedness as a characteristic of slavery, but he knew from his own experience that laws and company policies could not prevent enslaved people from seeking their freedom. Sometimes the pursuit of freedom was quite audacious and in the open. In 1856, a Virginia newspaper reported that a white man "saw a free negro go to the ticket office, exhibit his free papers, and then purchase a passage ticket to a neighboring town." The free Black man then gave the ticket to an enslaved person, who promptly departed on the train. "If impositions of this sort can be practiced on railroad companies by free negroes," the paper concluded, "owners of slaves have no earthly security for their property."[61] Likewise, a St. Louis newspaper complained in 1854 that enslaved people were "boldly" escaping "in our midst." The blame lay at the feet of steam transit, and the newspaper worried that such "wholesale plunder will prove destructive to slave property in St. Louis and the adjoining river counties, unless steps of the most extraordinary kind are taken to prevent it."[62] But as numerous accounts of escape attest, these laws and corporate policies were never fully successful.

Steam transit offered an enticing opportunity for enslaved people. Whether encased in a box or hiding in plain sight as a passenger, enslaved people made use of steam transit as the quickest possible way to put maximum distance between themselves and their owners. The penalties if caught could be severe, but the abundance of attempted escapes during the antebellum era attest that enslaved people saw the opportunity and seized it; the risks were worth the possibility of freedom.[63] They were active in demanding their freedom, and steam transit offered a powerful incentive and means to claim that freedom. Black people escaping slavery made use of whatever aspects of steam transit they could. When they could not purchase tickets, they jumped aboard, creating their own way by cobbling together different pieces of transit routes. In so doing, Black people continued the precommodified travel that predated steam transit. If using steam transit in the commodified way was too dangerous, enslaved people found another way to put the same infrastructure to use.

Perhaps the most extreme example of using steam infrastructure was when enslaved people would attempt to send themselves as freight. This required willing accomplices on both ends: one to send the box without raising suspicion, and someone to receive it knowing the value of the contents. The most famous instance of such an attempt was the case of Henry Box Brown, who accomplished the feat in 1849.[64] But he was not the only person to try this method, nor was he the first. In one such example from 1845, a box on a Memphis wharf was "left by a free black who was very particular in directing it to be handled with care." After the box had been there for an hour or so, someone on the wharf heard "a voice from the interior" say "Open the door." When the box was opened, "out jumped a strapping negro, nearly dead with suffocation and steaming like the escape pipe of a steamboat." The enslaved man revealed that "the intention was to ship him, in the manner attempted, to Cincinnati, whence he was to be conveyed to Canada."[65] In another instance, an enslaved man was transported in a box on a steamboat in 1848 from Charleston, South Carolina. The box was addressed to someone in Philadelphia. But after exhausting his food supply, he broke out of the box and was discovered when the boat alighted near Newcastle, Delaware.[66]

In 1854, a group of fifteen to twenty Black people traveled from Illinoistown (modern-day East St. Louis), Illinois, to Keokuk, Iowa, encased in "boxes marked as goods," from whence they went to Wisconsin.[67] In yet another case, from 1849, the *North Star* reported that "two large square boxes" were delivered to an express company in Richmond, Virginia, and suspicions were aroused when someone loading the boxes onto the railroad "heard a sort of grunt" from one of the boxes. A single man was found in each box once they were opened. The white man who had boxed them up was present when the unboxing happened, and he immediately went to board a train leaving for the North. A warning was sent by telegraph ahead of the train, and he was arrested at Fredericksburg and brought back to Richmond—thus demonstrating the danger to accomplices as well as those attempting escape.[68]

Historian Cheryl Janifer Laroche describes another pair of escapes via boxing up the enslaved person, one in 1854 and 1859, both of which made use of a regular run of steam packet ships on the Chesapeake and Delaware Canal. This steamboat line allowed these enslaved people to escape from Baltimore to Philadelphia.[69] Finally, in 1860, an enslaved man, Aleck, attempted shipment from Nashville to Seymour, Indiana, a fourteen-hour journey. When at one stop along the way, the crate in which he was stored was "rather unceremoniously thrown onto the platform," causing it to break apart, and

Aleck was exposed. The box in which he attempted his escape was displayed at the express company's office in Nashville, "where men and boys took turns climbing into it and local wags tried their hand at doggerel to commemorate the event."[70] In these extreme examples, taking advantage of freight shipment allowed for concealment throughout the entire route. There was always danger of suffocation or lack of food and water, but for these men and women it was worth the risk.

Far more common, however, were enslaved people who took trains or steamboats without resorting to packaging themselves up. Antebellum newspapers were full of stories of attempted escapes. Perhaps the most dangerous—because of the higher risk to the enslaved person and the explicit challenge to the color line—were the enslaved people who attempted to escape by passing as white. In other cases, a sympathetic white person might claim a Black person as a servant, while in reality aiding their escape. Such escape attempts were made by hiding in plain sight. In 1836, a train conductor in Washington, DC, grew suspicious of a white woman who said a Black woman was accompanying her as a servant. It turned out that the white woman was hoping to use this ruse to take the enslaved woman north and secure her freedom.[71] A pair of enslaved people attempted a similar escape in Mississippi; the one who could pass sat with the white passengers, and the other sat in the "negro car." While the latter was sitting there, "a negro train hand passed through the negro car, and recognized our black passenger seated therein as an old friend." The man protested that he belonged to the man in the other car, but the employee had them both arrested when the train arrived at its next stop. The two enslaved men were jailed.[72] In 1849, the *North Star* published the story of Ellen and William Craft, who successfully escaped through the use of steam transit. They chose to escape by openly posing as a white owner (Ellen) and her slave (William). By traveling openly, they decided, they could take "the most rapid conveyances," highlighting the appeal of steam. The ruse was successful, and they traveled "*one thousand miles in four days and a half, through the enemy's country.*"[73]

A few years later, another couple attempted the same escape with disguises. The husband, "quite dark and of small stature, was disguised in female apparel, and passed as the servant of his wife, who is white." Along the route, they encountered people who, not knowing the true identities of the couple, gave the wife advice specifically about how to travel with an enslaved servant. While on a steamboat on the Ohio River, the "servant" was told several times that "she" could break for freedom when the boat arrived in Cincinnati, but she demurred that "she liked her mistress very well, and did not care to leave her." The white "owner," for her part, received an offer from the boat's captain to land at Covington, "on the Kentucky side, where she could keep her slave safely, while she remained at Cincinnati.—This offer she politely declined, saying that she trusted to the fidelity of her servant." But when the boat did arrive in Cincinnati, both made haste farther North. "The cars on the underground railroad *do* travel fast," the article concluded.[74] This couple let white assumptions—such as those about a "loyal servant"—assist in their escape.

Other escapees also used white assumptions to their advantage. In 1852, an enslaved man named Robert rode a steamboat from New Orleans and got as far as Memphis before being discovered. Upon learning his true identity, the other passengers were "shocked because the fugitive had mixed easily with them and eaten regularly at the captain's table." He was "well dressed, well spoken, and well mannered," and had "darkish skin that may have made his dinner companions think he was of Spanish descent."[75] Robert's actions led others to assume things about him that were not true, and he had no reason to dissuade them. In 1856, Charlotte Giles and Harriet Eglin donned "billowing black clothes and the heavy veils of mourning" and, so disguised, took the train north from Baltimore to freedom.[76] These women counted on the deference and sympathy due mourning women to avoid questioning. In 1858, Mary Millburn boarded a steamship in Norfolk, Virginia, dressed as a man and managed to fool the authorities, since she was safely taken to Philadelphia. Millburn knew men could travel more freely and without questioning. For her and others, hiding in plain sight was a risk that paid off.[77]

While previous examples illustrate "open" escapes, "hidden" escapes were facilitated by steam transit as well. Enslaved people who made this attempt could not take advantage of the commodified travel that steam transit offered, but rather they needed to create their own routes to freedom. In 1838, an anonymous slave narrative was published that outlined how brutally enslaved people were treated while working to build the Hamburg and Charleston Railroad. Since the laborers were hired out, the overseers "did not care whether we got well or not," but they whipped the workers indiscriminately. After one particularly bad whipping, the author resolved to leave, and the railroad he had been building provided the perfect opportunity. He hid himself among some cotton bales on a train and then leaped off the train before it arrived in Charleston. He eventually made his way to Boston, hiding on a boat.[78] Steam transit changed the nature of travel for white travelers, who no longer had to create their own routes or find food and lodging where they could. By contrast, when enslaved people used steam transit, their use retained a precommodified character. They jumped on and off of transportation in order to remain unseen and had to depend on the sympathy of others in order to make it to their destination.

For example, enslaved people sometimes could count on the assistance of steam transit employees. When one fugitive boarded a train in 1853 but had no money for the fare, the conductor's "heart was touched, and soon the poor victim of oppression was stowed away in one corner of the baggage car, and was rolling on his way to Canada." But the support that this enslaved person received from employees did not stop there. When "a well-dressed Kentuckian" boarded the train to seek the man, the conductor instructed the enslaved man to jump off at the next stop and hide until the train began again, at which point the enslaved man should "run with all his might for the woods." The enslaved man did so. When the train started again, the master spotted the fugitive and demanded that the conductor stop the train. The conductor refused, saying that "they were already behind time, &c." When the slaveowner offered a hundred dollars for the train to stop, the conductor still refused. The slaveowner then made the engineer a similar offer, but "the engineer understood the game, and on went the cars with increased speed, at the rate of forty miles the hour, to the next dépôt, some thirty miles distant, all hands enjoying the torment of the slaveholder."[79] Enslaved people risked much when they encountered strangers on transit, but if the right person was compassionate it could make all the difference.

Concerned railroad employees had the ability to aid fugitives. In 1855, a newspaper article alluded to the fact that the New Albany and Salem Railroad, in Indiana, was apparently allowing fugitives to escape with the connivance of railroad employees. The accusation was that "a man who has charge of the scales in the railway depot at New Albany, is constantly giving aid to runaway slaves by hiding them under the scales until the moment the cars are ready to start, when, according to preconcerted arrangement with those on the cars, they are brought from their retreat, safely placed on board, and sent on their way rejoicing." In response, a railroad official protested that any employee providing such assistance "would, I am sure, be instantly discharged from the company's service." The newspaper article concluded with a warning: "The State of Indiana has five railways running from the Ohio river in a northernly direction, all of which are more or less liable to this same trouble about fugitive slaves. They are the roads commencing respectively at Evansville, New Albany, Jeffersonville, Madison, and Laurenceburg. The Jeffersonville road is so managed as to give no aid to runaways, and the result of the present troubles will, no doubt, be to place the New Albany road in the same category."[80] The varying response here—a sympathetic employee, an unsympathetic boss—highlights the challenges that fugitives faced when seeking assistance. Some employees were kind and others were not, and it could be impossible to know which was which until after making contact.

The experience of Ned Davis, who encountered unfriendly employees, illustrates this danger. In 1854, Davis attempted escape by lodging himself on the outside of a steamship, the *Keystone State*, which was departing Savannah. As the boat went on its journey, he was repeatedly dunked into the water—a very different travel experience from those who purchased tickets. Eventually he was discovered, quite soaked and exhausted, by the crew. Davis was jailed so that he could be returned to the slaveowner in Georgia. A New Jersey newspaper declared that Davis's "willingness to encounter so great a risk to obtain liberty, is a wholesome condemnation of the slavery institution."[81] And Davis did not give up after he was returned the first time. The following year, he "succeeded in fastening himself under one of the cars" of a train and was discovered only when a conductor on the train noticed that "the bottom of one of the cars cast an unusual shadow" from his lantern.[82] Once again, the fact that travel had been commodified made little difference to Davis's journey. Davis encountered a conductor who was unsympathetic and foiled Davis's plot. And Davis clearly risked significant personal harm in order to claim his freedom.

Other callous railroad officials were present in Sandusky, Ohio, in 1860. Two enslaved families were captured by slave catchers and placed in a darkened railroad car on a siding—a track holding cars separate from the main track. A crowd had been raised up to secure the families' release, but it was unable to locate the families due to their being hidden in the darkened car. The families were taken back to Kentucky and into slavery. After the families had been taken back, the train's conductor revealed to the press, that "someone he took to be a police officer from Cincinnati" had requested that he take aboard the train "six or eight passengers, 'mail robbers and counterfeiters.'" The conductor did not find this to be "a particularly unusual request," and so he agreed. The railroad superintendent added a special car to accommodate the request. When the conductor went back to collect the fares from the passengers, however, he then realized that he had been duped. Only then did he learn that the families were Black. As historian R. J. M. Blackett notes, "The darkened and locked railroad car suggests that the railroad superintendent was complicit in the abduction of the families."[83] It also suggests that those wishing to abduct Black people in the North went to some lengths, by this time, to avoid the observation of the public.

For those who were able to make it across the Ohio River or the Mason-Dixon Line, the development of steam transit unquestionably hastened the journey between those points and Canada. Historian Robert Churchill has noted that through the 1840s certain Underground Railroad routes in New York "diminished in importance . . . as rail connections in the region multiplied." In a similar vein, railroad expansion throughout the 1840s and 1850s in the Midwest had a profound effect on the ability of enslaved people to move quickly once they crossed the Ohio. "By the 1850s," according to Churchill, "a fugitive boarding a train in Philadelphia, Harrisburg, or Cincinnati could reach the safety" of the region most sympathetic to fugitives "within twenty-four hours."

While not universally the case, it is true that once in the North fugitives had the opportunity to switch from stealthily adapting steam transit networks to their own ends to participating in the commodified travel that was available to the wider white culture. This fact is demonstrated by the fact that the growth of steam travel changed the work that some abolitionist activists had to do. Rather than figure out how to house people as they moved on foot from one house to another, activists began spending time raising money to purchase transportation tickets. Churchill argues that in some regions of the North, "the operations of the Underground Railroad evolved more in response to technology than in response to the law." Indeed, "Underground networks that had been vibrant in the 1840s found themselves bypassed by rail transportation in the 1850s, and their operations gradually tailed off. With the opening of the rail link between Pittsburgh and Cleveland, most fugitives elected to travel by rail rather than along the old overland network running up the Pennsylvania–Ohio border."[84] Steam transit in the North presented a safer, more reliable, and swifter option for enslaved people seeking freedom.

For those who successfully made it to Canada, steam transit could be the last leg of the journey. One writer in 1837 recounted seeing an escapee "waive her handkerchief, from the deck of the Steamer" as she crossed from the United States to Canada, bidding farewell to the country that had enslaved her.[85] In 1850, the *Albany Evening Journal* published a note from Toronto that said, "Fugitive slaves arrive here by almost every steamer, from the American side."[86] In an 1852 tale of escape, a steamer captain chose to get wood and water "on the Canada side" of the Detroit River, "thus enabling the fugitives to avoid the risk of a recapture at Detroit."[87] While the Underground Railroad's reputation was for being "underground," in fact, aboveground transportation was a critical factor in getting fugitives to freedom—both railroads and steamboats. Steam changed mobility and opportunity for all people but did so in different ways. For white people, steam travel was commodified; it offered a sense of adventure and travel opportunities. For enslaved people, steam travel offered opportunity for escape, but steam travel retained its precommodified character.

Steam Transit and Abolition

Northern abolitionists were acutely aware of steam's power, both metaphorically and in the real world. Steam proved to be a valuable metaphor for abolitionists, as we have already seen with the musical piece "Get Off the Track!" In 1838, Lydia Maria Child used a steam metaphor to suggest relentless effort: "What think you of the prospects of Anti Slavery, dear brother? Shall we keep the steam up till the work is done?"[88] Another antislavery writer turned to the metaphor of a steamboat to illustrate the staggering human cost of slavery: "If all the innocent blood which have been drawn from the backs and veins of our race by the whites were collected into one reservoir, a steamboat might sail upon it."[89] In 1851, *Frederick Douglass' Paper* criticized churches that did not speak out against the Fugitive Slave Act by referring to steam transit's impact on the body: "As when we are on the rail-cars and boats, no one feels the motion, but each thinks the world is moving the other way; so those who are in these great bodies, will continue to recede with the retrograde-motion of the body, unconscious to themselves both of their motion and of their real position."[90] Steam transit could lull the senses into misunderstanding the true nature of movement, and so too were churches lulled into misjudging the necessity of fighting slavery.

We have already seen how the railroad was used as a metaphor for progress, and the Black press joined in this type of speech as it advocated for abolitionism and freedom. Here the "progress" of technological improvement was linked to the "progress" of eradicating slavery from the United States. A Black newspaper commented in 1841, "The Liberty car is a very safe one on which to jump and ride."[91] In 1852, the *National Era* wrote that freedom was "travelling with railroad speed."[92] The following year the same newspaper published an even longer metaphor: "The car of reform, which has been toiling up the inclined plane for the last five years, has acquired velocity. The crowd of scoffers which have heretofore lined the track have disappeared, and scores of passengers are jumping on—*thinking* men, with callous hands, are putting their broad muscular shoulders to the wheels."[93] In this metaphor, the abolition movement was literally a *movement*, and it became appealing enough that no one would want to be left behind. The railroad metaphor could also be applied to slavery itself, although in some cases the prospect of unceasing movement was more menacing. Referring to an upcoming election, the *National Era* warned about a bad electoral result: "The bloody car of Slavery will roll on triumphantly for another quadrennial term, dragging its wretched victims in its train, and trampling upon the rights and crushing out the spirit of Freedom."[94] Both positive and negative steam transit metaphors abounded in the antebellum era. Speed and momentum could mean progress and modernity or threaten relentless continuation of the status quo.

Perhaps the most common and telling metaphor of all was that of the Underground Railroad, which gave the image of a direct route to freedom. As Eric Foner has argued, the Underground Railroad was actually a "quasi-public institution," with its activities widely reported in the local press.[95] And as we have already seen, the Underground Railroad made substantial use of steam transit, which also put its work in the open. When describing their work to the public, abolitionists gleefully extended the metaphor, issuing "reports" similar in language and tone to actual annual reports by railroad corporations. In 1853, *Frederick Douglass' Paper* mimicked the language of the railroad corporate report when it commented, "The Underground Railroad was never in a more hopeful condition. The Fugitive law has raised the stock on some of our Western tracks, at least 50 to 75 per cent. Some new tracks have lately gone into successful operation, and the old tracks have undergone a thorough repair."[96] The paper printed something similar the following year, observing that

> the Underground Railroad Company is doing a very large business at this time, and the stock is up above any other Company here. We have had, within the last ten or fifteen days, fifty-three first class passengers landed at this point, by the Express train from the South. We expect ten more tonight. They all look well. I think our conductors take first rate care of them on the way. We have not had a single disaster on our Road, though our trains run at night altogether. We never want any headlights; our engines all know the depots, and turn off places without lights.[97]

In this instance, the paper made use of people's familiarity with machinery and corporate reports to highlight the liberating railroad to freedom.

Outside of metaphor, Black abolitionists and their white allies were acutely aware of the power steam transit had to liberate enslaved people. In 1840, the *Colored American* made a very explicit argument about the link of railroads to freedom. The paper noted that just a few years prior, someone escaping slavery might need a week to walk from Baltimore to New York and would constantly be in danger of capture. By contrast, "Now so extensive are our railroads, and such the arrangements, one leaving upon the arrival of another, that a poor fugitive, may leave Baltimore in the morning, and the third night following, may find himself safely in Canada, a British subject." The writer concluded that railroads could have no "better purpose" than helping people escape slavery. "May the *railroad mania* in our country increase," declared the author, underlining the value that steam transit had for abolitionism.[98] The same newspaper voiced similar sentiments in the following year, declaring, "Abolitionists ought to be the friends of internal improvements, even though it should bankrupt the nation, while it would assist in freeing the slave."[99] Surely no white southerners intended that steam transit networks would ease the escape from slavery, but that was one of the most significant unintended consequences of construction.

The routinized use of steam transit to assist people in escaping slavery is apparent in Sydney Howard Gay's "Record of Fugitives." Gay kept a log of the fugitives who arrived at his office in New York City. His notes make clear the multifaceted roles that steam played in assisting escape. In 1855, for example, this log revealed that enslaved people who approached him for assistance had variously "arrd here by Steamer"; escaped with the assistance of a train baggage master; escaped on a steamer during a "stormy night"; traveled on the railroad from Baltimore to Philadelphia; walked over twenty miles and, until finding a train station, used railroad tracks to guide their path north; paid a steamboat steward $60—"all he had"—to be concealed on board; or was concealed on a steamboat by a captain.[100] These diverse uses show how enslaved people used steam infrastructure in a variety of ways for their own ends. Instead of purchasing tickets, they bribed stewards or relied on the assistance of sympathetic baggage masters and captains. Enslaved people had to navigate the steam networks in very different ways from white passengers. Gay's log makes clear that enslaved people were ingenious and persistent in making use of steam transit.

But if steam could aid enslaved people in making their way north, it could equally bring southern slaveowners north in pursuit of what they claimed as their property. As historian Charles Bolton has noted, "white Americans viewed running away as disobedience rather than as crime. The masters of fugitives were responsible for getting them back, but states and local governments provided them with assistance."[101] While slaveowners never hesitated to reclaim what they saw as their rightful property, the Fugitive Slave Act of 1850 gave additional bite to their demands.[102] In a sign of how tenaciously white southerners fought for slavery, they would venture north to reclaim enslaved people who had broken free. In so doing, steam transit was critical to their efforts, and transportation termini—like wharves or depots—could become sites of contestation and protest when word spread about kidnapping and the local population was sympathetic. Thanks to the decentralized nature of the enforcement of fugitive slave laws, which resulted in white Americans taking matters into their own hands, steam transit across the country became implicated in how enslaved people fled and how white Americans worked to take them back into enslavement.

While the Fugitive Slave Act codified the ability of slaveholders to reclaim Black people, those efforts far predated the law. Even worse, steam transit could be used to capture free Black people in the North and spirit them away to slavery. In 1836, for example, five men were taken on board a steamboat and "shipped to the South as fugitive slaves. As no persons have been arraigned before the city courts as fugitives, it is supposed that these persons... were literally kidnapped and hurried off to the South without the form of trial." The *Liberator* wrote about this case and warned, "Human tigers walk abroad in New York, with impunity," ready to capture Black men and send them South.[103] The ready availability of steam transit made it possible for such "tigers" to claim their prey and move quickly before protest could be roused. In 1837, the Cincinnati *Philanthropist* reprinted an item that warned, "We have frequently heard that there are steamboats that make a practice, particularly at wood yards, of raising quarrels with negroes and kidnapping them, for the southern market."[104] In 1838, a free person of color from Rochester, Jason Verplank, had been jailed in Louisville, Kentucky, after being removed from a steamboat "on suspicion of being a slave." Another woman on the same boat, Betsey Green, jumped overboard to avoid the same fate, and drowned.[105]

A similarly frightening story came from the *North Star* in 1849, which published an account of several free Black people who suddenly realized that they were on a train about to enter Maryland and were thus in danger of being sold into slavery if they were kidnapped. Several of the free Black people jumped from the train. When a woman was about to jump, the writer, fearing for the woman's safety, prevented her from doing so. But the woman was insistent: she "implored us to let her go, that she would be arrested and sold into slavery; she fell on her knees and wrung her hands—a more painful or affecting sight we never beheld." The writer asked an official for the railroad to intervene, and the official "had her provided for and sent back as soon as the train from Baltimore arrived."[106] The sequence underlined for the writer the horror of being sold into slavery and once again demonstrated that steam transit was not just a rapid means to freedom but could also lead back to slavery.

Steam transit was a public means of transportation, and public spaces like wharves and train stations became battlefields if fugitives were discovered. Depending on time, geography, and political leanings, crowds could favor or oppose the fugitive's freedom. Accounts from the time highlight these tensions. One attempted capture in Ohio in 1852 illustrates how the moment of transferring from one transportation mode to another could serve as the time at which capture was attempted. When a group of seven Black people moved from a railroad to a steamer, they were accosted by a group of southerners attempting to return them to the South, assisted by the city marshal. During the ensuing confusion, two of the seven "fled and probably escaped on some of the steamboats that were just leaving the dock." When the others were arrested, a group of local free Black people immediately began to protest. It was discovered that the city marshal did not have the authority to make the arrest, and "instantly, the colored people who had gathered round, hustled the fugitives out of the room and down stairs," and they boarded a sailboat, "the evening steamboats having previously left." The southerners "were unable to find a sailor in the city, to their honor be it said, who would go himself or permit his vessel or boat to be used in pursuit."[107] The public nature of steam transit presented opportunities both for kidnapping and for public intervention to prevent it.

There were many other such cases in the antebellum era that highlight how technological infrastructure became sites of protest about the institution of slavery. In 1854, Black people in Harrisburg, Pennsylvania, gathered around a train that carried "a colored man ... supposed to be a fugitive from Kent county, Maryland." When the crowd could not prevent the train from leaving, "they stoned the cars, without, however, doing any serious damage."[108] When another fugitive was captured and jailed by a US marshal in Racine, Wisconsin, that same year, one hundred men arrived via steamboat and demanded the man's release. When that was refused, they broke into the jail and spirited him away.[109]

According to historian Robert Churchill, crowds grew in what he terms the "Contested Region" of the United States as opposition mounted in that region to the Fugitive Slave Act and the techniques of law enforcement. In the late 1850s, crowds gathered in Columbus, Zanesville, and Iberia, Ohio, at different times to protest the carrying off of fugitives. As Churchill notes, "the collapse of public support for enforcement of the law is best illustrated by the behavior of federal marshals in the late 1850s. When called upon to arrest fugitives in the Contested Region, these officers did their best to perform the arrest as quickly and quietly as possible." In the case of the arrest in Columbus, federal officials "lured John Tyler to the train depot in Columbus, seized him despite the resistance of bystanders, and hustled him aboard the train to Cincinnati." A crowd attempted to free him at the Xenia, Ohio, depot, "but the train departed before he could be taken off." The officials successfully spirited Tyler out of Ohio and into Kentucky.[110] In this region, shifting opinions about slavery were borne out in the public reaction to captured runaways in public places. Transportation, which by its nature required doing work in the open, thus presented an opportunity for those who opposed slavery.

Large crowds in a public place could turn violent. When slave catchers arrested James Worthington in 1854 in Akron, Ohio, "a large interracial crowd soon gathered." When the slave catchers threatened violence, the crowd became "infuriated . . . and one bystander warned the slave catchers that 'if they dared to exhibit arms, they should be torn to pieces.' As the crowd closed in, the slave catchers let Worthington go and slowly backed away into a waiting train." Another "interracial crowd of several hundred men, armed with muskets, pistols, and clubs, assembled at the depot" in Limaville, Ohio, in an attempt to rescue another enslaved person being carried back to the South. The assembled crowd had "not blocked the track, however, and the train put on steam and sped through the station without stopping. An observer acknowledged that the rescue had been poorly planned, but dared those involved in the rendition to 'try it again.'"[111]

The issue of Black people being returned to the South was a controversial political topic in the antebellum era, particularly after the passage of the Fugitive Slave Act of 1850. As slave catchers moved north in pursuit of fugitives, transportation links were vital in speeding them back to slavery. Depots and wharves were public areas, and as sites of transfer for slave catchers, they became sites of public protest and confrontation. Crowds could gather at a depot to demand the release of a fugitive. The fact that railroads had to stick to a schedule—regardless of who was on board—no doubt heightened the tension of these confrontations. In some cases, the crowds were successful. In other cases, slave catchers had the cooperation of the railroad companies, employing special cars and side tracks to hide their true purpose from the community. Although there was a national law governing fugitives, enforcement and community reaction could vary by time and geography. The efforts of enslaved people to secure their freedom show that every trip taken on steam transit was fraught with the possibility of getting caught and sent back to slavery.

Liberating Those Who Traveled North

In addition to southerners coming north to reclaim their "property," there were also times when white southerners brought enslaved people with them when they traveled north, and points of transit became moments when enslaved people could find their freedom. White southerners were aware of the danger of bringing enslaved people to the North. As historian Jon Sterngass has observed, one "South Carolina commentator noted that the possibility that one might actually sit 'next to a runaway negro' in a railway car was enough to cool the ardor of any self-respecting planter who wished to visit Northern watering places."[112] According to historian Edlie Wong, slaveholders would "select attendants with kinship ties in slavery as a guarantee of their continued compliance once in a free state." Thus, travel to the North did not necessarily represent an opportunity but, as Frederick Douglass had argued, would be a "threat." Enslaved people had to "weigh their emancipatory desires against separation from all personal ties."[113] The cost of freedom was stark if it meant leaving family behind.

In 1847, the *Alexandria Gazette* republished an item from a Buffalo newspaper, no doubt attempting to illustrate for its readers in Virginia how dangerous things could be in the North. As the article recounted, an Alabama slaveholder traveled to Niagara Falls and brought an enslaved woman with him. As they sat on a train ready for it to depart, they found that "between twenty and thirty colored persons, as was previously arranged among themselves, rushed to the cars and attempted to take by force the object of their sympathies—some throwing obstructions on the track while others mounted the cars. They were resisted by the conductor, engineer and others.—A general *melee* ensued—stones and brickbats were freely used, by which a number were seriously injured, when the train finally got under weigh."[114] The crowd made its own effort to rescue the enslaved person, and since this account was in a Virginia newspaper, the crowd's actions were presented in a threatening way. The story showed that Black people in the North were not afraid to take matters into their own hands to free an enslaved person, and the point of transit formed a place at which they could fight for that person's freedom. The anti-slavery *Emancipator* gave a slightly different account of the same incident: "The violence was on the other side." The newspaper quoted someone as saying that a "colored man" asked the enslaved woman "if she wished to go back to the South. He was knocked down. A row followed, in which several colored men were very much injured. Finally the Southerner and the woman left in the cars, and she has probably returned to slavery."[115] The bravery of the free Black people in this case was considerable, and the fact that this was done in the public venue of a rail car made the request all the more dramatic.

A Salem, Ohio, case in 1854 attracted a great deal of attention from the newspapers. An anti-slavery meeting was occurring in Salem when word was received (via telegraph) that an enslaved girl would be passing through the town on the railroad. The meeting moved to the depot, and a crowd of about twelve hundred people gathered there to hear an antislavery speech by Charles C. Burleigh. According to eyewitness Henry Blackwell, a committee was appointed to board the train, speak to the conductor, and approach the young girl, apparently about eight years old, about whether or not she wished to be free. The girl replied affirmatively. Blackwell, "seeing that the passengers in the cars seemed to sympathize with the owners," assisted in getting the girl off of the train. There was an "angry discussion" with the owners, but eventually "the bell rang, the cars started & the assembled multitude gave 9 hearty cheers."[116] In this case, the train's bell signaled the end of the debate and the beginning of freedom for this young girl. The need for the train to keep its schedule cut off the debate.

The cases recounted so far demonstrate that confrontations on steam transit could have very different outcomes: in some cases the crowd liberated the enslaved person, and in others they could not prevent a return to slavery. The uncertainty doubtless caused some fugitives to take extra precautions. In 1854 in Syracuse, a group of abolitionists received notice that an escaped fugitive, bound by US Marshals, would be on an approaching train. A crowd of two thousand people assembled and met the train when it arrived. There was no fugitive on board, but one newspaper account noted that a free Black man on board "fled, but was soon overtaken and the matter explained." Perhaps he feared that the assembled crowd was not there for his safety but to capture him and transport him to the South. This episode and that person's reasonable fears highlight the uncertainty for Black people and their mobility during this extraordinary time.[117]

Yet, in an instant, a person's life could change thanks to the opportunities afforded by steam. In 1855, an enslaved girl named Rosette Armistead (or Rosetta Armstead) began speaking to a free Black man while on a train near Cincinnati. The man told Armistead that "she had a right to her freedom, inasmuch as she had come into the State by consent of her master, or his agent." According to the newspaper account, she could hardly believe her good fortune. Armistead asked, "Are we in a free State?" and after receiving the affirmative response, "at once said she wished to be free." When the train stopped, Armistead disembarked, and the residents of the town succeeded in getting a writ of habeas corpus and secured her freedom.[118] Steam transit made a crucial difference in Armistead's life, as it allowed a chance encounter with a stranger who was both bold and kind.

Discrimination in the North

While travel in the South was governed by the laws of slavery, travel in the North was also fraught for Black people, who faced discrimination when they traveled.[119] As historian Elizabeth Pryor has noted, "vehicles of public transportation—stagecoaches, steamships, and railroads—emerged as one of the most notorious places for antiblack aggression." In part this was because travel was a public act: Black people were publicly claiming spaces mostly inhabited by white people. Additionally, to return to Douglass's statement at the beginning of the chapter, mobility was tied to freedom. To be free was to be able to move about without needing another's permission. Thus, there were undoubtedly white northerners who saw Black mobility as a threat. "To assert white authority in these spaces," Pryor concludes, "stagecoach drivers, steamship captains, railroad conductors, and white passengers threatened, insulted, and forcibly ousted colored travelers to insist that access was white-only domain."[120] Employees and passengers alike enforced a whites-only concept of public space. The presumed "need" for enforcement underlined that the treatment of Black people on northern transport could be capricious and uneven. While laws and company rules abounded, there was still amazement expressed at the time at how different treatment could be. In 1841, the *Massachusetts Spy* wrote that the Eastern Railroad in that state had separate cars for free Black people. But if a southern slaveholder wanted to bring an enslaved person into the first-class car, "not a word will be uttered against the arrangement. After all, then it is *not* color alone which excludes a man from their best cars. The colored person to be excluded must also be *free!!*"[121] Therefore, it was not necessarily just the color of skin that truly made the difference in where people sat in the train.

The injustice done to Black travelers in the North was a common theme for the editors of the Black press. In 1859, Frederick Douglass wrote, "The negro is the test of American civilization, American statesmanship, American refinement, and American Christianity. Put him in a railcar, in a hotel, in a church, and you can easily tell how far those around him have got from barbarism towards a true Christian civilization."[122] When a Black person was out in public, as on steam transit, the treatment they received from their neighbors illustrated the level of commitment those neighbors had to equality. Train conductor Charles George recalled in his memoir, "The aristocratic southerner, who had no idea of objecting to the presence of a negro valet or nurse in the same seat with himself, or wife, or children, would not tolerate the same negro in the car if he were traveling as an equal on a ticket he had purchased for himself."[123] Enslaved, Black Americans could travel with white Americans, but once they demanded equal treatment, Black Americans could suffer at the hands of their fellow passengers or corporate employees.

Despite this unpredictability, editors in the Black press encouraged their readers to travel by steam. Charles Ray, for example, was the editor of the *Colored American* and encouraged emigration to Wisconsin by saying that "persons can step on board a steamer at the foot of Barclay street, in this city, and they need not step on land again, until they arrive at Green Bay, in Wisconsin, unless they choose."[124] But Ray also expected equality when traveling, writing, "Is not A MAN, A MAN, whatever be his complexion? Our brethren should DEMAND the treatment of FREEMEN, or cease traveling about the country." The offense done to Black passengers enraged the author: "Sooner would we walk from New York to Philadelphia and back again, than submit to one tenth part of the insult and proscription which many of our brethren travel under."[125] For Ray, it was galling to see low-class white people admitted to steam transit when he was barred. "Why should a colored man, who is equal in wealth, in education, in refinement and in taste," he asked, be "crowded into . . . dog-cars on the Railroad, or pantries on board the Steam-boats?"[126]

Historians have argued that the Black press encouraged a middle-class morality and respectability among its readership in the antebellum era.[127] Ray embraced this argument, believing that his respectability should mean that he be treated as well as any other man of his class. "A white 'black-leg' can travel in our best cars, and have the accommodations of our best steam-boat cabins, with their cards, their guns, and their dogs," he pointed out in 1838, "but a colored gentleman, of education, wealth, and piety, whatever may be the comforts and refinements of his home, cannot visit his friends abroad, without being degraded and insulted." Eventually, Ray came to believe that it was better to shun steam transit altogether than to submit to ill treatment—thus, he concluded that piece by arguing that Black people should "hold intercourse through the Mail, or go on foot."[128] In 1840 he wrote that "some colored people will not eat on board of steamboats unless they can do it upon the principle of equal rights."[129] Ray urged Black people to boycott, but it appears that most Black passengers suffered indignities in order to continue to travel.

The treatment of Black passengers was maddeningly inconsistent, with passengers not knowing necessarily how they would be treated until they got on board. There, they were subject to the whims of those who controlled the space. In September 1837, Ray recounted his experience on a steamboat departing New York. He was denied tea by the ship's staff, being told that he would have to take it in the kitchen. Ray and his companion rejected this different treatment and spoke to the boat's owner, who was on board and who informed them that "this is the general practice." Ray defied the owner, noting that he had been served on other boats without incident—illustrating the inconsistency of treatment. The owner then consulted the man who was "acting as captain," who also then refused service, finally saying that "the waiters do not like to set a second table." Ray's conclusion was that each person "palmed off" responsibility on the other, "clearly showing it to be a principle of which they are all ashamed." Notably, Ray and his companion refused taking tea in the kitchen, telling the boat's staff, "We do not like to be the agents of our own degradation."[130] In so doing, Ray demonstrated not only his fearlessness but his demand to be treated with dignity. Steam transit was place of public display, and Ray demanded that others treat him with dignity in public.

In 1838, David Ruggles complained of the treatment he suffered on a railroad. Once again, the reception he received on board the train differed from what he expected when he bought his ticket, demonstrating inconsistency within the company. The conductor refused to seat him in the car he had paid for but instead called Ruggles "a d—d [damned] abolitionist" and in conjunction with other passengers "forcibly ejected me from the car, and forced me into what they call the pauper (or jim crow car)."[131] Clearly he had been sold a ticket, but once aboard he was subject to the whim of the conductor, aided by other passengers. Ruggles recounted the discrimination he suffered on both railroad and steamboat in the space of just one month in 1841. In June of that year, he was ejected from a railroad when he "refused to move to the seats reserved for people of color," and a few days later the captain of a steamboat would not sell him a more expensive ticket which would have allowed him to "avoid sitting in the windy forward deck." Two weeks later, he was again removed from a railroad car for rejecting instruction to sit in the segregated space. He brought a lawsuit against those who had thrown him out. He lost the suit but gained widespread approval in the abolitionist press.[132] Ruggles continued to battle against discrimination in the North, but his experiences illustrated how conductors and captains bore the power to enforce racial laws. The enforcement was also carried out in the full view of the public, causing humiliation.

Jeremiah Myers, a Black man, detailed in 1848 how the enforcement of conductors could turn violent. He was allowed to purchase a first-class ticket on the Stonington Railroad, but when he attempted to claim his seat, the conductor "threw him from the platform, endangering his life."[133] Another Black man, minister Henry Highland Garnet, told a similar story about the Niagara and Buffalo Railroad that same year. When a conductor attempted to eject him from the railroad, "I quietly returned towards my seat, when I was prevented by the conductor, who seized me violently by the throat, and choked me severely. I have been for many years a cripple. I made no resistance further than was necessary to save myself from injury; but nevertheless, this conductor and another person, whose name I do not know, continued to choke and to assault me with the first."[134] These stories demonstrate the fickleness of discrimination—Myers had been sold a first-class ticket, so clearly at least one company employee felt he belonged there—but also the harsh violence that was required to enforce discrimination.

Company employees would enforce discrimination even when it defied logic to do so. Frederick Douglass noted that he was poorly treated on a steamboat in 1852 and blamed the steamboat company, pointing out the capriciousness of the treatment. He was told that he could not purchase a place in the steamboat's cabin "because white passengers would object to his presence in the cabin." But that made no sense to Douglass, since those passengers "were the same that had rode side by side with him by railway, without the slightest objection, or manifestation of dissatisfaction." The author blamed "the officious clerk, proud of his brief authority, anxious to appear superior to somebody on earth," who "took especial pleasure in asserting his superiority over a negro."[135] Nothing about the passengers had changed in the transition from railroad to steamboat, but the clerk on the boat took the moment to boost his own authority.

Not all steam transit companies were equally equipped to segregate passengers. One author noted that when passengers were not segregated on a particular train, there were no objections. "As there is on that train but one class of cars," the author wrote, he was able to sit "among white and among colored persons, as the case may be. There is no other alternative, and we rode along, no one raising an objection, nor apparently thinking of any, nor exhibiting signs of one."[136] Such treatment demonstrated that the decision to separate passengers based on race was ultimately arbitrary: if this railroad company could keep different races together in the same cars, then surely others could do it as well. Similar thoughts came from a writer in the *Colored American* in 1840, who reported that during a recent steamboat ride he received "all the privileges of the boat" except being able to take meals with the white passengers. Nevertheless, the author noted that when he had ridden a steamboat eight years prior, he had been "ordered from an humble place on the promenade deck, to the forward one, and there only allowed to remain." The author believed that there was "a softened and improved state of feeling, a higher regard for the rights, and a greater respect for the feelings of men, than formerly existed."[137]

In other cases, Black passengers were apparently left alone. An anonymous traveler reported on a steamboat ride in 1836 that "of the passengers from N York to Philadelphia there was a full proportion of colored to the white compared to the population of the state. Several of them in dress and appearance, and in every thing but color equal to the whites. And though none of the latter would associate with them, they appeared to enjoy themselves quite as well and to take a pride in showing that they were as well off."[138] In this case, the white passengers did not want to mix with the Black passengers, but no one complained to the corporation demanding that they be removed from the boat. The Black passengers were allowed to remain on board and enjoy the trip, hoping to demonstrate that they were as respectable as any other white passenger on board.

Not all passengers were as understanding. Ejections from steam transit were at the whim of conductors or captains, but they also included the complicity of the other passengers on board. In 1841, a writer reported his experience on the New York and Harlem Railroad. The writer took a seat in the cars, noting that he "was as well dressed as any in the car, the conductor not excepted." The conductor approached him and said that he must stand outside the car; he could not take a seat. When the writer protested that he was sick, the conductor responded, "Sick or well, you must stand outside or get out" and noted that "he had his orders to let no colored person ride inside." The writer hoped that other passengers in the car would advocate for him since he was sick, but "every one of them said, by their silence, CRUCIFY," and so he disembarked. The author concluded, "When the white public comes, it matters not how ragged, how dirty, or how drunk, they can be accommodated. But the colored, clean or dirty, sick or well, must stand outside, else not at all. Shame! shame!"[139] Segregation was not a simple matter of rules; it also required the collusion, even silently, of others on the train or steamboat. We have seen that crowds could turn the tide by expressing a common opinion, such as when rescuing an enslaved person brought to the North. But equally so, the silence of white passengers could serve to stabilize discrimination.[140]

Beyond supporting the acts of conductors and captains, passengers would enforce discrimination in other ways. The *National Era* published an account in 1854 of a man on a train who beat two Black passengers in order to make room for himself and a friend. After the Black passengers had been forced from their seats, the white friend remarked, "You gave him a lesson—I did not notice they were blacks, until you spoke!" The writer lamented, "All this, Mr. Editor, passed within the eyesight of at least thirty persons; yet no one said a word in *defence*" of the Black passengers, but rather approved of what happened.[141] This illustrated the complicity of the public in the treatment of Black passengers as well as the fact that differences among people were not as obvious as racial ideology prescribed ("I did not notice they were blacks") but rather required violent enforcement.

Another white passenger attempted to preserve racial hierarchy in 1855, when J. W. Loguen boarded a train and asked if an empty seat was occupied. "Yes it is," came the response, which the newspaper characterized as "Lie No. 1." The exchange continued: "'Where is the gentleman who claims it?' 'He has stepped out.' (Lie No. 2.) 'Then I will take it until he returns.' 'You will have to give it up pretty quick.' (Lie No. 3.) 'Oh very well, I will do so.'" Loguen's neighbor grew more and more agitated at having to sit next to him and eventually left. As he walked off, Loguen called after him: *"I'm afraid if you go away I sha'nt know the man whose seat I've got when he comes!"* and the rest of car gave Loguen's opponent a "laugh of derision."[142] Here, public expression was on Loguen's side. Could Loguen sense that the crowd in the train was backing him, and thus he could risk the confrontation? The history of discrimination on steam transit suggests that there was no guarantee that public reaction would be supportive.

In short, for Black Americans the inconsistency of treatment proved to be one of the most exhausting aspects of steam transit. At each leg of the journey, Black travelers would have to fight for dignity and respect again and again. On one journey, they might encounter difficulties as Ray and Ruggles documented above. And yet another journey might pass without incident. Ray wrote in 1838 that he could travel for "17 hours, from New York to Boston, by steamboat and by railroad—no one uttering a word of wonder, or casting a look of surprise, that I should participate equally in the privileges, the enjoyments, and the reflections by water and by land, of which it was the right of others, on board, to partake."[143] There was no easy way for a Black traveler to know what type of journey he or she might have. They could receive the kind treatment that Ray described in his 1838 trip, or they could be stopped at every turn by racist conductors, captains, or other passengers.

During the era of steam transit, Black mobility was a topic of considerable debate, legislation, and newspaper reporting. For Frederick Douglass, one of the things that defined an enslaved person was rootedness to place, enforced by a white master. If enslaved people moved about, it was because of the master's will. While Douglass identified an enormous barrier to Black mobility, it is also true that enslaved people took matters into their own hands and used steam transit to escape, breaking free of the master's hold on their mobility. Enslaved people were critical to building the transportation infrastructure of the South. In so doing, they built the means that others would use to flee slavery. Free Black people also worked on wharves and in steamboats, and in mingling with enslaved people they could offer advice and opportunities for escape. While there was a national law for fugitive slaves, its enforcement varied across time and space. Attempts to return enslaved people from the North to the South led to public confrontations, and as steam transportation infrastructure developed through the antebellum era, railroad depots and steamboat wharves became public sites where abstract debates about liberty, freedom, and slavery were instantaneously made acutely real. As enslaved people were transferred from one transportation mode to another in the North, there was an opportunity for local crowds to decide that they would not tolerate the actions of slaveholders in their own territory. The expansion of steam transit infrastructure heightened the tension around the issue of futigives, as transit routes proved popular means of escape. While steam transit had commodified travel for white passengers, enslaved people making use of steam for escape fell back on an older, self-designed method of travel where they had to patch together their own routes, hide rather than risk paying fares and being discovered, and rely on the kindness of strangers. Even in the North, the ability of Black people to travel by steam remained contested. Some Black passengers were able to travel unmolested, while others were roughly handled by employees or other passengers. Treatment could be arbitrary or impulsive, as when Black passengers were sold tickets but then denied services once they boarded. The question of equality of treatment would not be solved in the antebellum era, but the roots of those issues were planted early in the era of steam transit.

CHAPTER SIX

White Women Passengers

In 1806, Sara Chester Smith confided in a letter that she "felt a great deal of reluctance to crowding myself in-to a public stage, with a company of men and but one of my own sex."[1] In this statement, Smith succinctly identified several of the worries that some white antebellum women had when considering travel. Travel in a stage could be unpleasant, since it required close quarters with strangers. And for women, travel was potentially fraught because those close quarters would be with unfamiliar men. Alone on a stage—or with "but one of my own sex"—Smith was concerned about how she would be treated by others on board. Within a few years after Smith's letter, the development of steam transit would open up a range of travel opportunities for her and other white women, far more than what was available by stage. Yet while the modes of transit may have changed, for these women the fundamental questions about the propriety of travel and its potential dangers remained the same. Was it appropriate to travel alone, or was a chaperone necessary? New modes of travel also created new questions: how should one behave on a steamboat or train? Was it appropriate to converse with strangers? What practical considerations should women take into account when contemplating travel by steam? This chapter addresses how white women confronted these questions.

Many of these questions revolved around how white women should comport themselves within the particular *spaces* of steam transit—a railroad depot, a steamboat deck, and so on. Steam transit came about as Americans were wrestling with the changing meanings of *public* and *private* and how those spaces were specifically gendered. In her study *Separated by Their Sex*, historian Mary Beth Norton argues that in seventeenth-century America, "notions of rigid, gendered divisions between the terms *public* (male) and *private* (female) did not exist."[2] This division began to emerge in the mid-eighteenth century, where public and private were tracked onto separate genders, and private, more "specifically . . . came to be equated with *domestic*."[3] By the time that steam transit became popular, this distinction was more settled in the public eye. Historian Mary Ryan has shown how women had "secured cultural representation" in public celebrations in the 1840s and 1850s, but they were there mostly as "prescriptions for domesticity." Such public displays were an attempt to fortify the connection of woman/private/domestic. While women were linked to domesticity and privacy, public spaces were open to men, particularly in the realm of electoral politics.[4] The meaning of public and private, and its gender associations, had shifted from colonial times, and that terrain would continue to shift throughout the nineteenth century.

Space on steam transit did not sit neatly in the public/private dichotomy. With a few exceptions for state-owned works, all of these spaces were owned by corporations. Thus, as private property they were not expressly "public" spaces as a town square might be.[5] At the same time, railroads and steamboats were open to many travelers—and thus were not completely "private" spaces, either. Over time, different classes of transit developed, but even the wealthiest passengers could not fully avoid rubbing shoulders with others. Historian Amy Richter has argued that railroads functioned as multiple spaces at once: as "a commercial space subject to the demands of the market, as a mobile space carrying people beyond local controls and knowledge, as a small and intimate space challenging notions of what constituted respectable contact among strangers, as socially diverse and fluid space capable of blurring the lines of class and caste." Travelers moved back and forth across these spaces while traveling.

As we will see, corporations attempted to create some physically separate spaces for women that reinforced the private/domestic association: a separate car on a train, perhaps appointed with more comfortable seating as one might find in their home parlor. Thus, a woman traveler faced the challenge of "maintain[ing] her respectability as she traveled back and forth among an almost domestic ideal of privacy, an expanded network of polite society, and, finally, a public world of social differences, new experiences, risks, and possibilities."[6] Richter effectively details the challenges that women faced in the postbellum era, and the roots of this challenge lay in the antebellum era, as the evidence herein illustrates. Space on antebellum steam transit was complex with many spaces intertwined, and women navigated these public and private worlds within a larger world that expected women to behave in certain ways in public and in private. With the boundaries between such spaces more fluid on steam transit, some women took the opportunity to break out of the spaces prescribed to them.

Through an examination of the experiences of white women, we can see how American culture reflected back on steam transit; the interchange of steam transit and culture was not unidirectional. Strictly speaking, there was no *operational* reason for trains or steamboats to accommodate women in a particular way. Adding a separate car for women did not make a train go faster. But the introduction of women passengers meant that steam transit entered a broader cultural conversation about private/public distinctions and how women should behave and were perceived when outside the home. Steam transit had an impact on American culture, and the experience of white women shows how culture, in turn, impacted steam transit.

New Opportunities

There were some who felt that women should not travel by steam. Massachusetts farmer Asa Sheldon railed against women's mobility in a speech to the Farmer's Club of North Reading. In the speech, he complained that "some young ladies spend a great deal of their time riding in the cars." Sheldon argued that "if they would spend less of their time in this way, and more of it in assisting their mothers in that portion of farm work belonging to the female sex, they would be healthier, happier, and in the married life more contented with their husbands."[7] Sheldon identified that steam transit allowed women to break out of prescribed domestic/private roles.

Needless to say, women did not take Sheldon's advice, and corporations did not assume they would. While transit corporations did not break down their ridership by gender—making it impossible for the historian to analyze statistics—the sheer volume of women writing about travel and the abundant advice literature on travel makes it clear that women were enthusiastic about the possibilities steam transit presented. There were many reasons to travel. Steam transit brought the opportunity to strengthen social bonds. In 1815, one woman wrote to a friend that a steamboat company had started up nearby, and she was glad that this would "be an easy and pleasant way for us to break in upon your circle when the season is advanced."[8] Steam would make it more convenient to visit friends, and the writer here made explicit reference to the social season—the time for "calling" on friends. Inquisitiveness could be enough. Recall from chapter 2 that Martha Parker rode a newly constructed railroad out of curiosity. Women's education gave women plenty of opportunities to cultivate a spirit of inquiry. As historian Mary Kelley has demonstrated, the subject of geography was "one of the staples" in women's education of the antebellum era, a time when more and more women received formal education.[9] Surely for some students, learning geography must have made the prospect of travel enticing when presented with a wealth of new opportunities via steam transit.

Corporations responded to this growth in women's travel by seeking to accommodate them.[10] The open design of American steamboats and railroad cars allowed for plenty of intermingling—by race, class, and gender. But companies worked to offer accommodations to women, both in stations and en route, and thereby give them a space separate from men. They did so to encourage ridership, promote safety, and comport with societal norms that matched women and domesticity. In 1837 the Boston and Worcester Railroad's directors elected to change the passenger house at Worcester so as to "furnish a more suitable place for the ladies."[11] In 1840, the Western Railroad's board of directors "ordered, that the Engineer & Agent provide two cars with exclusive accommodations for ladies, either by altering those now in use or procuring others."[12] That same year, the periodical *Yankee Farmer* spoke glowingly of ladies' accommodations on the Norwich and Worcester Railroad. A car had one "apartment" for ladies and their gentlemen escorts, and another apartment that was "solely" for ladies. The latter was "carpeted, and in every respect beautifully furnished, with wide and convenient sofas, dressing table and mirror, wash stand, and other arrangements for the comfort of passengers."[13] In 1848, the president of the Louisa Railroad (of Virginia) wrote to a car manufacturer, requesting that the "Ladies Apartment" have seats which were "red plush sofas."[14] Another railroad company's ladies' cabins also featured "luxurious sofas for seats and . . . a washstand and other conveniences."[15] And some train stations included separate waiting rooms for men and women.[16]

A variety of factors drove the creation of these separate spaces for women. Part of the compulsion was doubtless paternalistic: an attempt to provide a space for "poor" women who were "compelled" to travel alone. Parents and husbands may have also been comforted to know that their daughters and wives would be comfortable and have space separate from the men. Such spaces could cut down on overt sexual harassment or abuse—or errant tobacco spit—from men. By allowing women to mix solely with other women, these separate spaces could also assist women in navigating the multi-tiered space that Richter described, since women could converse with each other and nervous travelers could learn from more experienced women. Ladies' cabins in trains and steamboats were areas for women into which unmarried men could not enter—or at least were not supposed to. Richter notes that ladies' cars were generally open to men chaperoning women or husbands traveling with wives.[17] As historian Walter Johnson has argued, such spaces "spatially reinforced the idea that white women in the cabin were virtuous rather than promiscuous, no matter how far they were from home."[18] Furnishings in such areas for women attempted to mimic the private space of a home parlor. Corporations went to lengths, therefore, to accommodate women on their journeys and in so doing re-created the domestic spaces that were seen to be a woman's proper place.

Such actions were part of a wider movement to create a built environment for women in the nineteenth century.[19] Hotels, for example, were working to attract women with "ladies' entrances, parlors, and dining rooms."[20] Nevertheless, the project of creating a separate built environment was not universal or complete. Not all corporations put up the expense of separate accommodations in every car, steamboat, and depot. And these spaces were not open to women of all classes. Women who traveled on the main deck of a steamboat could only get privacy if they hid behind the cargo, but those in the cabins could be safe from "unwelcome or harmful contact with obnoxious philanderers."[21] There remained plenty of times when travelers had to mix. We saw in chapter 2 how Elizabeth Koren praised the comfort of the steamboat she was on but despaired over the "mixed company" on board.[22] The antebellum era thus represented a time when corporations began to accommodate women, but the infrastructure was not always such that women could retreat to their own space. Thus, for many white women the crucial consideration was whether or not they should travel alone.

Traveling Alone

Should women travel alone? Male chaperones could play an important role in travel. They fended off prying strangers, ensured that the luggage made it to the right destination, secured food at station stops, and attended to other matters. By taking on these tasks, chaperones made women's experience more private—men could take care of the interactions with conductor, baggage handler, ticket seller, and others. A chaperone's actions helped the woman maintain the fiction that she was in a private space by removing all—or at least most—public interaction. Such women traveled as if in a small bubble of privacy, removed from others. Chaperones could perform this service even if they did not know the woman. Sometimes a chaperone would be an acquaintance of another man in the woman's life, and if that man considered the chaperone to be trustworthy, that was sufficient. Therefore, the chaperone could be a total stranger to the woman being protected. Some women chafed at this restriction or found it odd that a stranger's care could ever be preferable to none at all. But other women clearly valued traveling under the protection of a man.

The position of etiquette manuals on women traveling alone was mixed. Surveying the entire nineteenth century, not just the antebellum era, historian John Kasson found that "although female readers of etiquette books were frequently congratulated on the comparative freedom they enjoyed in America to venture forth in the city without male escorts, increasingly toward the end of the century they were warned 'not to abuse it,' and to 'err on the side of caution rather than... of boldness.'"[23] Kasson's finding suggests that as the century progressed women were less likely to travel alone, perhaps because of societal expectations. By contrast, C. Dallett Hemphill argues that "the conduct literature tends to corroborate those contemporaries who insisted that American women traveled without fear of molestation."[24] Patricia Cline Cohen has noted that women made a tradeoff when they traveled with a chaperone. They may have gained respectability in the eyes of the general traveling public, but they also suffered the "denial of female competence in negotiating public space."[25] My own reading of the evidence, including the letters and diary entries of women travelers, is that regardless of their ultimate choice, the choice was made thoughtfully, and antebellum women were fully aware of the implications of selecting one over the other.

Some writers cautioned women of catastrophic consequences of traveling alone.[26] In an extreme example of this type of literature, Warren Burton wrote a broadside in 1848 called "Moral Dangers of the City," in reference to Boston. He warned that women were subject to being recruited as prostitutes when they entered the city. This could happen because men preyed on women "under pretense of securing for her respectable lodgings, or a situation at service," and the men would instead "conduct her to a house of infamy." Burton directly tied the danger to steam transit: "Lounging pimps" stationed themselves at "railroad depots and steamboat wharves" in order to "beguile the bewildered and inquiring girl to the brothel."[27] In 1858, another book, titled *Tricks and Traps of Seducers*, also forewarned women that traveling without a protector could lead to prostitution. The stories in the book described alarming tales. In one, a fourteen-year-old girl was tricked into prostitution in New York City. Another similar story in the same book was included expressly "to show the rashness, the almost wickedness, of sending young girls on a journey without a protector."[28]

Such tales of exploitation represented one extreme of advice offered to women, that traveling alone might lead to severe danger. Given the fact that women did travel alone, such warnings may have served to titillate the audience more than to dissuade women from traveling. Indeed, there was plenty of advice and growing infrastructure for women who wished to travel alone. As we have already seen, steam transit companies created specific spaces for women travelers in an effort to provide them comfort and safety. Etiquette writers gave more specific tips to women traveling alone. In 1853, *Godey's Lady's Book* published a set of advice under the heading "The 'Unprotected Female.'" The article opened with the acknowledgment that traveling alone could be a "frightful undertaking," but it continued that "it seems to us perfectly proper, so that a lady conducts herself with all due reserve and decorum; and, especially in our own country, there are very few 'lions in the way.'"

The article then gave a series of tips for independent women traveling: do not take too much baggage, do depend on the train conductor for recommendations on where to stay, and do not take strangers into confidence. Of course, the conductor may well have been a stranger, but his position of responsibility presumably made him trustworthy.[29] A manual from 1853 urged women to consider that if she was wanted to be "treated with respect" while traveling alone, then "her own deportment must in all things be quiet, modest and retiring."[30] In 1860, an etiquette manual instructed women on what to do "if you travel under the escort of a gentleman" rather than make the assumption that women *must* travel with a man. This word choice reflected that women themselves had options, and the etiquette guides could help women make either choice.[31]

Godey's Lady's Book had recommended relying on the conductor as a trusted adviser if the woman traveler knew no other men on board. And some railroad corporations instructed their employees to give special care to women passengers, recognizing that some would be traveling alone. In 1850, the Michigan Central Railroad instructed station masters to "pay particular attention to the wants and convenience of the passengers—especially ladies, in the care of their baggage, direction to other conveyances, &c., and in all their connexion with the public, be civil and accommodating." For their part, passenger conductors were told to "attend ... to obtaining proper seats for ladies, and to a proper disposition of their baggage" prior to the departure of a train. When a train arrived, they were to "not leave their passengers until the whole of the baggage is distributed, aiding in its distribution, and generally attending to the wants of their passengers—especially those of the ladies."[32] In 1855, the Eastern Railroad instructed its conductors to "see that all passengers, especially ladies, are properly accommodated."[33] A sermon preached at the funeral of a railroad employee in 1856 noted, "Women now travel in the cars, without friends, and sometimes to great distances. They naturally look to conductors for protection and advice. The conductor should be a true gentleman."[34] From corporate regulations to praise from ministers, such writings suggest that employees were expected to watch for women passengers, and that the position of conductor carried with it expectations of helpfulness and trust.

What of the women themselves—what did they make of these options? We know from women's writings that there were plenty who preferred to travel with a chaperone. Maria Dorthea Furman was relieved after locating an escort for her trip in South Carolina. The lack of a chaperone was the "one great impediment" she faced before traveling, she wrote in a letter. For Furman, "it would be highly improper as well as extremely unpleasant to undertake any thing like a journey without the protection of a gentleman." She reiterated the point in another letter a few weeks later, in which she wrote, "It is altogether impossible for us to even go a few miles without a gentleman, strangers as we are to the country, and entirely ignorant of the roads." Had she not been able to obtain male oversight, "we should have been obliged to remain stationery."[35] Furman, then, perceived that the lack of an escort meant that she could not travel at all. Likewise, Maria Bryan Connell worked hard to find an escort for a train that departed at midnight. She wrote, "I have tried my best to persuade Uncle to go with me but he will not consent." She talked to several men, but all were unable to accompany her. The result was that she was simply unable to travel: "No one here ever thinks I need any recreation or pleasure, more than a door post," she complained. "I cannot think of any future time that I can go, for the difficulties will be no less hereafter than now that I see."[36] Connell clearly chafed at staying home—she lacked "recreation or pleasure"—but also decided she could not go out without a chaperone.

Women traveled with chaperones they knew as well as those they did not. Strangers could be socially acceptable over the course of a journey as long as each could vouch for the next man's propriety. For example, a woman's father or husband might accompany her to the train station, ask around to find a suitable person, and place the woman under the stranger's care. Known to the woman or not, the function was the same: the chaperone's presence announced to the public that the woman was under care. An anonymous author in the *National Era* newspaper observed this in practice and described it for the readers. The author encountered a "pretty Yankee girl" while traveling, which allowed an opportunity to "tes[t] the truth of the remark of foreigners upon the respectful treatment which unprotected women receive in America." The writer observed the girl as she journeyed. Although she was under a different man's care "with almost every change of the boat and cars," the writer found that "in every instance was she placed in charge of the most respectable, gentlemanly, person present; and their attentions could not have been more respectful and delicate, had she been a mother or a sister" to the chaperone in question.[37] If, according to convention, each gentleman could vouch for the next, then a woman traveler could have safe passage while traveling. When men vouched for each other, there was public recognition that each man in the chain was capable of providing protection. This public affirmation guaranteed the respectability of the women under protection.

At the same time, letters and diaries of antebellum women suggest that there were plenty who were unafraid to travel without male protection and even delighted in doing so. Mary Lyon, founder of Mount Holyoke College and advocate for education, underwent a shift in her own thinking. She took a railroad journey in 1833, and wrote to a friend that the journey was unpleasant in part because she was alone: "I had no protector, no one to take care of my baggage, and engage a hack after I should arrive." She would have preferred having someone along who could manage the business of traveling for her. But later in life, she challenged those who criticized her for traveling so much: "What do I that is wrong? . . . I ride in the stage coach or cars without an escort. Other ladies do the same."[38] She had become comfortable traveling alone and did so routinely. Sarah Waller described traveling without a male escort in 1848 and found sufficient female company to keep herself entertained. When she arrived at Cincinnati on a boat in the evening, she "remained on board all night as there were other ladies remaining." The following day she suggested to the other ladies that they venture out and visit the "Greek Slave," a sculpture by Hiram Powers, which was then touring the country. They went out "without a gallant"—that is, a male protector—and found the sculpture "without difficulty."[39] Waller and her friends demonstrated that women could travel, enjoy the sights of new locations, and experience new things without the protection of a man.

Other women were quite content with an independent travel experience. In 1839, Lucretia Mott, another active social reformer, wrote to a friend that once she was not traveling alone but had heard of a man who was concerned that she was, and she retorted, "And what if I was!"[40] In 1854, Caroline Barrett White got assistance from a man getting her luggage checked with the "necessary receipts & checks," but then traveled alone in the train and she "<u>whirled</u> & <u>steamed</u> along to Worcester." The next month, she received assistance from a man in labeling her trunks, again traveled alone in the train, and at the transfer "got my baggage all aboard & checked" without anyone else's help.[41] In 1855, Elizabeth Cady Stanton mused that if women dressed as men did, then they "could travel by land or sea; go through all the streets and lanes of our cities and towns by night and day, without a protector; get seven hundred dollars a year for teaching, instead of two or three, as we now do."[42] For Stanton, this different treatment—and salary—only stemmed from outside appearance. If Americans could see beyond these veneers, women could have significantly more freedom.

When traveling alone, women had to confront the issue of whether or not it was appropriate to talk to strangers. In the etiquette literature, both men and women were allowed to talk to strangers, and it certainly does not seem like travelers were reluctant to talk.[43] The fact that fellow travelers were temporary strangers "may have heightened the confidences shared by female passengers."[44] Indeed, Rebecca Russell Lowell Gardner found that on an 1831 steamboat journey, other women "all seem'd eager to unbosom themselves to someone, & each came up to me voluntarily while I was making arrangements for the night."[45] Perhaps it was easier to speak frankly to a stranger knowing that the relationship was not permanent.

If speaking to a stranger was acceptable while passengers were en route, etiquette guides frowned upon continuing any familiarity after the journey was over. A book in 1838 informed women that if a man introduced himself to a woman in a "proper and respectful manner," then women could respond "with politeness, ease, and dignity." But "acquaintanceships thus formed must cease where they began, and your entering into a conversation with a lady or gentleman in a boat or a coach does not give any of you a right to after recognition."[46] Travel thrust people together in tight spaces and conversation could help pass the time, but there could be no expectation that "friendships" would extend beyond the trip itself. People whom one met while traveling, another guide declared, "have no claim to more than a passing bow if you afterwards find that the acquaintanceship is not particularly desirable."[47] A manual from 1853 noted that "incidental acts of politeness should always be acknowledged with thanks; but they should not be construed into a desire of commencing an acquaintance."[48] Politeness on steam transit was important, but it should not extend into a longer relationship once the voyage was over. The temporary nature of the acquaintance provided an easy "out" if the strangers' company proved distasteful.

The decision whether or not to travel alone was a key decision faced by antebellum white women. Women were acutely aware of the importance of this decision and the judgment that it brought upon them from the eyes of other travelers as they moved through railroads and steamboats. For some, traveling under the watchful eye of a chaperone was key—they may have lamented the difficulty of finding a male protector but could not imagine being without one. For others, traveling alone was not troubling. Much of the concern here was about appearances and comportment. Such concerns about women's behavior and its effect on men were loaded with cultural concerns about sexual propriety. Therefore, the choice that women made to travel alone or with a protector was not an idle one, and antebellum women understood the stakes of their choices.

Etiquette Guides

All women—those who traveled alone and those who did not—could turn to the copious etiquette literature of the antebellum era to learn how to behave on board. Guides to etiquette have a long history in the United States and predated steam travel.[49] As more and more people left their hometowns in the early nineteenth century, it became necessary to provide guidance so that people would know how to act in a "world of strangers" and in situations they had not previously encountered. Etiquette guides—primarily but not exclusively targeting women—entered the conversation to address these concerns. There were a substantial number of etiquette books published during this period, and many included sections on travel. Patricia Cline Cohen counted "some seventy" etiquette guides "published between 1830 and 1860." The frequent level of publication suggests that there was some demand for this material in guiding women on how to acquit themselves.[50] Although there were many books published during the time period, that does not mean that there was a wide variety of opinion on etiquette topics. Indeed, a quick perusal of these books reveals that they featured not just "similarity of materials but frequently a wholesale repetition of passages." Lax copyright enforcement led to readers frequently seeing the same material over and over again. Nevertheless, as steam transit grew more important in American life, these books expanded their contents to include guidance for those venturing out into the world about how they should act on railroads and steamboats.[51]

Etiquette guides contained a whole range of advice on comportment, from personal behavior to social courtesies. These guides reminded their readers that their actions would be judged by other passengers. Therefore, readers had to be on constant alert and monitor their actions. The topic of "body carriage" received wide coverage: "scratching oneself; drumming the table; biting one's nails; fingering one's nose or ears, excessive coughing and yawning, and so on."[52] Such advice, long provided by etiquette guides, was ripe for adaptation to circumstances on steam transit, when people were thrust into public display. This public display was not simply about appearances and nail-biting; the potential for judgment ran much deeper. Authors of etiquette guides argued that traveling was an excellent time to assess a person's true character. It was all very well to be a paragon of politeness when sitting in a parlor or when invited to a dance, but the stress and uncertainty of travel presented a truer test. Then could people genuinely judge how polite a person was: not when they were in comfortable society, but when being polite would actually inconvenience them, or when being polite benefited a stranger of unknown background rather than a fellow member of their social class.

Authors recognized that behaviors when traveling needed to be regulated in order to ensure that people got along in harmony in what could be trying circumstances. "If in travelling, the duties of politeness are less numerous," one such etiquette book noted in 1833, "they are not, therefore, the less obligatory."[53] In 1856, another etiquette guide chided its readers: "Under no circumstances is courtesy more urgently demanded, or rudeness more frequently displayed, than in traveling. The infelicities and vexations which so often attend a journey seem to call out all the latent selfishness of one's nature; and the commonest observances of politeness are, we are sorry to say, sometimes neglected."[54] When traveling, people left their familiar surroundings, but this author pointed out that the frustrations of travel were no excuse for poor behavior.

Why bother with politeness? One 1836 etiquette book noted that politeness was a "common bond" that would help passengers get along when put in close quarters. The author wrote that "every real lady" had a "duty . . . to set the best examples in manners and conduct to the crowd around her." What did this specifically mean? A lady should never be "pushing her way rudely" through a crowd, "never seizing on a chair that another wishes for; never standing in the way; never staring at what is going on near her; never, in short, forgetting the convenience of others, but always calling forth their best feelings by treating them generously and courteously." Women could never forget that they were on public display and others would notice their actions. The author lamented that mealtime was when the demands of politeness were most often forgotten. She noted that she had herself gone without dinner on a steamboat "because I could not bring myself to contend for a place, or scramble for a seat, as all around me were doing." She concluded, "Who would not rather lose a dinner, than their self-respect?"[55] Politeness, then, was a way to ensure that traveling remained pleasant for everyone.

Etiquette guides directed men to show deference to ladies. One author, for example, instructed men that if they wanted to open or shut the window in a railroad car, they should ask any ladies present before doing so.[56] In return, women were urged to receive that deference graciously. One guide praised Americans for the "courtesy shown to women" by male travelers, which was closer to the "character of a father's or a brother's protecting care," as opposed to that found in Europe, which is "more of the lover in it; and is far less safe, and less to be relied upon." At the same time, however, women needed to demonstrate that they appreciated the care shown to them. "Every little service rendered by a travelling companion," the author reminded readers, "should be either civilly declined, or courteously accepted."[57] Similarly, another author warned of coldness toward those providing courtesy: "I have often seen men in steamboats, in stagecoaches, in churches, and other public meetings, rise, and give their seats to women, and the women seat themselves quietly, without a look or word of acknowledgement. . . . Avoid such discourtesy, my young friends—it is not only displeasing, but unjust." Travel could serve as a leveler, since all classes were mixed together, and politeness should also cut across classes. Women, therefore, were instructed to show gratitude toward all, regardless of social standing. The same author pointed out that women should not dismissively accept the politeness of a "plain, respectable man" when at the same time being effusive to "what *she* called a gentleman."[58] Courtesy when traveling should be shown to all regardless of social class. Such courtesy—freely offered by men and gratefully accepted by women—presumably eased burdens for those traveling alone.

To men's deference, women were expected (and instructed) to show gratitude throughout the antebellum era. One author in 1846 chided readers: "The best seat, my dear girls, is not your *right*, and should be accepted with some acknowledgement."[59] And in 1851, a writer celebrated that "nothing is more credible to our countrymen than the readiness with which they give up their places for the better accommodation of ladies," but the writer hoped that the ladies "should always render suitable acknowledgment when received."[60] An 1856 author worried that when he willingly gave up his seat, he received far more thanks from foreign women than American women. He remained hopeful that American women's behavior would change: "We believe that American ladies are as polite *at heart* as those of any other nation, but *they do not say it.*"[61] Women could expect good treatment on steam transit but were expected to reciprocate.

Whether or not men were adequately thanked, it seems that a general deference to women was followed in the antebellum United States. Irish immigrant Thomas Mooney wrote to his cousin back in Ireland that if a lady *"looks at a desirable seat, the gentleman possessing it rises by a sort of instinct to let her have it."*[62] George Templeton Strong complained about women's rights in his diary in 1853, claiming that political agitation was unnecessary, since "womanhood is still reverenced in this irreverent age and country, as every omnibus and railroad car can testify."[63] For Strong, the treatment of women on board steam transit was perfectly sufficient evidence that women did not need any stronger political rights. They were already well treated, and voluntarily so, by men in public, and that should have been sufficient for their needs, in his view.

Etiquette guide authors updated their works as the types of travel in the country expanded. The author of one such revised edition, in 1849, noted, "When this chapter on travelling was written, rail-road conveyance was unknown in the United States, and now it forms so large a portion of the traveller's experience, that I must add a few hints on the proper behavior of young ladies in rail-road cars and station-houses." The change in modes of travel warranted a reconsideration of the advice that was given. Fortunately, much behavior already known could be easily applied to the new types of travel: "What has already been said on the reserved and quiet demeanor necessary in public places, applies with great force to the thronged station-house and the crowded car." Nevertheless, some specifics were mentioned. Women were cautioned that they should not be "lounging and lolling . . . on the shoulders and in the arms of their companions." Particular care must be taken because the railroad was such a public place. While one might feel that they are surrounded by strangers, in reality those strangers could one day become friends, and if they witnessed ill behavior on the train, they would be "forever prejudiced ... in consequence of some indiscretion then committed."[64] Women could not afford to let down their guard when traveling; one never knew when a relationship with a stranger at a depot might change if the person was met in a different context. Etiquette manuals were ready to provide guidance on how best to behave and prepare women for all circumstances met along the way.

Practical Tips

Outside of advice on behavior, etiquette manuals had a wealth of other information for traveling women. This practical information addressed a range of topics. An obvious area for advice was luggage. One early guide urged women to consider the luggage they would carry with an eye toward not inconveniencing their fellow travelers. For example, women traveling for pleasure should not carry baskets larger than they could handle themselves, "or the burden of them will fall on every gentleman who waits upon them to or from a coach or steamboat." The guide here assumed that men would be at the ready to assist women travelers. By the same token, flimsy boxes should be avoided: "It is better to take a second trunk, than to plague everybody with the care of so frail a thing as a paper box."[65] Women should select luggage with a thought toward who would be helping them.

After selecting the right luggage, advice was given on how to pack. Women were urged to "keep those things on the top that you will need first, and when you are to set off early in the morning, to pack your trunk the night before, and leave out only such things as can be put in your carpet bag."[66] An 1851 guide drew parallels to other aspects of women's education: "The art of packing a trunk well is highly important and should be early learned. Those who have a correct eye for form and space or know how to draw, will learn it most readily."[67] Another guide illustrated the importance of packing one's own hairbrush through telling the story of a young traveler who asked the author's acquaintance to lend a hairbrush on a steamboat. The author's friend explained that brushes could not be lent, "for you know all persons' heads are not equally clean; and diseases have been communicated by persons using the same combs and brushes. It is therefore considered ill-bred to ask the loan of a brush." But the friend gave the young woman a brush to keep, which she graciously accepted, and presumably would not make the same faux pas again, having been so instructed.[68] After reading this story and seeing the young woman's embarrassment at having asked for a brush, the reader would hopefully avoid the same error.

Travel dress was an important consideration for women, and many etiquette guides offered advice in this category. Clothing had the power to signal—or disguise—one's social standing to others. Some thought had to be given to keeping clothing clean as well.[69] In 1836, one guide noted that "the plainer your dress is for travelling, the longer it will look nice."[70] Another guide from 1842 gave similar advice, noting that the clothing choices made by a traveler would not just determine comfort but also show signs to others about the wearer's breeding: "A plain dress is adapted to traveling, where you are liable to dust, and wear and tear of all kinds. Americans are reproached with overdressing in traveling, and are said to be marked in Europe by their costly traveling-dresses. . . . Certainly, artificial flowers, dangling ear-rings, gold chains, silk dresses, and French capes, are out of place in a steamer. Such unfitting articles should be rejected as vulgar, and as indicated a want of sense and *education* in the wearer."[71] Dress was not simply a matter of comfort but also sent an important signal to other passengers about the wearer's breeding. There was one item, however, that was valued for its use in travel, rather than appearance, and that was a watch. In 1860, a guide urged women to wear a watch, since it would help the lady manage the "one rule to be always observed in traveling," which was "punctuality."[72] We have seen how time management was pressed upon travelers in the antebellum era as critical to ensuring that they were able to make their travel arrangements. Women were not excluded from this important lesson, and thus they were encouraged to wear a watch.

Other smaller pieces of advice abounded, giving practical tips to the uninitiated in order to improve their travel experience. An etiquette guide from 1853 suggested that women tie a ribbon of the same color around each of their trunks in order to ease identification. The guide also recommended that women write down on a "large card" a description of each item of baggage, to be given to the man escorting them to aid him in knowing what baggage to look out for. The guide encouraged women to sit with their back to the engine, if they could, which would reduce the danger of flying sparks coming through the window. If this was not possible, the book offered several strategies for removing cinders from the eye. The book also discouraged reading, which was "very injurious to the eyes, from the quivering, tremulous motion it seems to communicate to the letters of the page. It is best to abstain from your book till you are transferred to the steamboat."[73] All of these tips combined to prepare women for a rail or steamboat journey.

Beyond the advice offered in printed guides, some men sent extensive directions to their wives or daughters in order to assist them in travel. In 1853, Lincoln Clark instructed his wife, Julia, on making a transfer in Philadelphia: "Follow the crowd directly along until you come to the companies omnibuses." He warned her, "Be careful where and how you step, and not to be squeezed by the crowd" when boarding a ferry, perhaps worrying that his wife would end up in the water rather than on the boat. Finally, he advised her to speak to the conductor about the best train to take once arriving in Baltimore, since "that will depend upon the train by which you will arrive."[74] In so doing, Clark identified the company's agent as a reliable source of information and one that could be trusted by a woman traveling alone. In 1854, Clark again advised Julia to check her baggage appropriately depending on which scenic sites she wished to see: "You must remember that the Falls and the Suspension Bridge are two different places."[75]

Other men wrote to their daughters with similar advice. In 1854, Salmon Chase wrote to his daughter that he regretted having left her at a train station alone (she was thirteen years old at the time). He confessed that "it was a blunder, as a good many of my acts are, and teaches the necessity of forecast even in small matters." Chase had neglected to consult the railway timetable before leaving the house, "the doing of which would not have taken three minutes." Had he done so, he would have realized that he could have easily traveled with his daughter to her destination and returned to Washington on an express train with plenty of time to spare. Therefore, asking her to travel alone had been unnecessary; he could have gone with her on the trip. Chase hoped, however, that his daughter would learn a lesson from his own mistake: "Don't procrastinate; keep your wits about you; and avoid harumscarumtivieness."[76] Likewise, in 1859, Lucian Barbour gave some advice to his teenaged daughter: "Take checks for your trunk and whatever else you may trust to the baggage car and keep your checks until you get your baggage. I shall be on hand if I know when you come. If none of us are at the depot, go directly to Mr Wilson's and trust yourself to no stranger."[77] Barbour's daughter was able to travel on her own, but her father insisted that she still take precautions, and he wanted her to know who was trustworthy and who was not.

Did these wives and daughters want this advice? Did they treasure the guidance or discard it as soon as they said their goodbyes? Were they building their own store of tips and tricks as they grew in experience, making their father's or husband's cautions increasingly tedious or unwelcome? The advice only comes to us from the perspective of the man who offered it: it may have been a welcome salve to anxious nerves or annoyingly aggressive micromanaging. Regardless of its use, the fact that it was given reemphasizes the importance of comportment in these new and complex public/private spaces. As women navigated these spaces, they accepted or rejected this advice as part of their own understanding of how to travel, and in so doing, women's travel experience transformed into something regular and natural.

Steam Transit in Women's Magazines

Women could read about travel in magazines targeted expressly toward an audience of women. This literature introduced steam travel to those who had not yet experienced it themselves, let women imagine themselves traveling, and overall helped naturalize travel for the readership. Women's magazines did not shy away from the topic of technology; "Let railroads be glorified!" proclaimed a story in *Godey's Lady's Book* in 1842.[78] Authors of fiction sometimes incorporated steam transit into their stories. Thus, steam travel could have meaning for readers even if they had never personally set foot in a railroad or steamboat themselves. Recall that the railroad seemed normal to Sallie McNeill in part because of her having read about it. Thanks to stories such as the ones considered here, readers like McNeill could become familiar with steam transit before they encountered it in person.

Steam transit was incorporated into storytelling in a variety of different ways: as a framing device, as the setting for a story, as a marker of modernity, or to indicate the distance of time between the reader and the story's setting. Travel could be a framing device for authors, allowing them to set the scene for their larger tale. In 1841, a story in *Godey's Lady's Book* opened with a conversation about travel. A young girl, who had "just come from boarding-school, her head filled with the romance of novel reading," complained about the state of travel in real life compared with that found in books. "Travelling is dull work," she exclaimed, charging that it required no "heroism" to travel by railroad or steamboat. To the amusement of the gathered adults, she complained that "there are no banditti to level their pistols at one's head; no highwaymen to demand your money or your life; no opportunity, in short, of exciting an interest in some dark-eyed fellow traveller, by requiring him to risk his life in one's defence." Her father then told her that "life has scenes more thrilling than were ever forged in the heated brain of a novel writer." He then continued that the travel held its own rewards: "For my own part,

I never enter a stage-coach or a steamboat, a railroad car or an omnibus, without finding something worthy of note among my companions—something that tells me of the hidden depths which lie beneath the dull surface of every-day life." The assembled group then asked the father to explain himself with examples, so the story could begin in earnest. In this case, travel was a framing device for the story: the daughter complained that the transition to steam was robbing travel of its romantic allure, and the father countered that any group of people, even on a routine railroad trip, could reveal interesting stories.[79]

Another story, titled "A Day in a Railroad Car," provides an example of how railroads were explicitly used as a setting. The story opened with the narrator musing that the railroad promoted a "social life" because "at each village there is a swarm of fresh passengers, and at each station a dispersion, and however brief their transit may be, there is some trifling intercommunication that discloses the condition and objects of the parties." Every stop offered the opportunity for new people to come on board and new opportunities for interaction and education. To the observant person, these observations could be of great interest, and they allowed the author to devise different ways to bring people together. At one part in the story, someone on the train points out that former president John Adams is a passenger. Later, the narrator observes the excitement with which the young children saw the train for the first time: "'Is this a car, grandpapa?' exclaimed one of the little girls scrambling over the sofas and chairs—'it seems more like a house.'" The girl's reaction reflected the image of domesticity that railroads attempted to create in outfitting their cars with furniture suitable for a parlor.

The story then captured a conversation between the grandfather and grandmother about safety. The lady urged her husband to choose the seat carefully: "My dear, you should select one of the last passenger cars, for I read all the railroad accidents, and they always escape; and you must sit about in the middle of the car to avoid some danger, I forget what it is—"; her husband then drolly responds, "And near a window to avoid some other." The wife, disconcerted by her husband's response, reminds him, "You remember that dreadful accident!" and he responds to reassure her: "Yes, I remember them all." In this (fictional) conversation, we can see how the author both raised concerns about accidents and smoothed them over. The couple was concerned about accidents, but the contradictory advice—sit next to a window to avoid one type of accident and far from the window to avoid another—led to a certain amount of resignation, and they simply carried on with their journey.[80] The author here reflected American acceptance of risk that was inherent in steam travel.

In other stories, the presence of steam transit was used to signal modernity. One author used railroads to demonstrate that people living distant from the city could have what was not possible before. In this story, railroads were characterized as "distance-destroyers," which allowed people in the countryside to keep up with fashion, and the writer observed that "the denizens of country-places ... are almost as much *au fait* with regard to dress as their city's neighbours."[81] Another author in 1846 distinguished between two characters by giving their opposing views on technology: "Mr. Hamilton was the advocate for railroads, steam engines, factories, education schemes, machinery, universal suffrage and vote by ballot; Mr. Saville of coach-traveling, hand-looms, restricted instruction and close boroughs."[82] The railroad was a clear marker of modern life.

Still other authors used steam transit to mark time in a story. In one story published in 1843 but set in earlier times, the author characterized the "interior counties of the state" as distant because they were "not then as now, accessible by railroads and canals."[83] The reader could easily imagine how the lack of steam transit would make it more difficult to travel to remote areas, and thus envision that earlier time. Another story from 1847 used the time required to travel as a sign that the story was taking place in the distant past: "it required many a day to perform a journey of a few hundred miles. It was, as I have said, a long time ago."[84] For the contemporary reader of that story, a journey of a hundred miles was quite easy to contemplate. And in a story from 1849, an author used the fact that a destination could only be reached by stagecoach as a symbol of its isolation. In this particular case, the character felt "certain that in these days of railroads a place must be comparatively secluded to which the readiest access was by a stage-coach."[85] The reader of this story would understand that taking a stagecoach meant that the destination would be well off the beaten path.

The purpose of much of this literature was to entertain. But the circumstances presented in the literature also had to seem realistic to the reader. Including steam transit as a setting immediately filled a story with a sense of possibility: where were the characters going? What sort of people would they meet? Would any accident befall them on the way? Alternatively, steam transit was a clue for the reader about the time in which a story took place. The emergence of railroads and steamboats marked a specific time in America, and to have a character refer to a difficult overland journey could help mark the story as taking place in the past, and help settle the reader into the time. If a place was only accessible by coach, it must be leading the story's characters to a sheltered—perhaps even mysterious—location.

In 1859, Gail Hamilton (the pen name of Mary Abigail Dodge) wrote a piece for the *National Era* in which she commented on the quality of her chaperones. By reading her assessment, we can see how women considered the advantages and disadvantages of both ways of travel. "Yes, it is very nice to travel alone," she mused. "Husbands, and fathers, and brothers, are very good in their way, and convenient and handy to pay your bills and look after your luggage, and we ought to be very kind, and considerate, and forbearing; still it is a relief, once in a while, to be free, to be your own man." Dodge's terminology here—to express the freedom of traveling alone as "be[ing] your own man"—demonstrates how engrained it was that traveling alone was tracked to a specific gender. Dodge could see the advantage of a chaperone but also wanted to break free from this stricture. She continued by recalling an earlier chaperone, who was a man "of whom I had never heard." Her experience was frustrating: despite the fact that he was a complete stranger, during the journey he had "charge of me when I wanted to take care of myself."

Dodge and the chaperone got along well enough, but the chaperone left the journey before she reached her final destination. Therefore, he "hunted up an old schoolmate, a planter from the South, and consigned me to him, ready invoiced and labelled!" Dodge felt herself treated like a piece of freight rather than a human being. The replacement chaperone was not impressive, and she skewered his methodical, plodding conversation and his knack for falling asleep. By the end of the essay, the tables had turned. He was jostled awake and asked, "Are you going to get out?" She responded, "Perhaps we would better, sir; the people seem to be getting out!" Thus *she* took on the role of the chaperone, paying close attention to the route and ensuring that they got out at the correct stop. Additionally, she "managed to pick up his goods for him, and land him safely at H——." Dodge fumed that she had to be seen under his care, "as if I couldn't take care of myself fifty thousand times better than that respectable stupidity could take care of me." Dodge was contemptuous of the care given by the men, and in her story poked fun at her lamentable series of "protectors" and demonstrated that she had, in fact, protected them.[86]

Dodge understood fully what role men were "supposed" to play in this situation. She gamely—or at least grudgingly—went along when she was accompanied by men. But her writing also shows that she was appalled with how she was treated. She did not need a man to make it safely to her destination, and even turned the tables on her supposed protector. Dodge's experience shows how steam travel could become familiar for women in antebellum America.

Women seized the expanded travel opportunities available with steam transit. By venturing out, women pushed against the strictures that confined them to domestic space. At the same time, travel took them into steam transit's complex public/private space. Corporations attempted to domesticate a portion of the travel experience by building separate spaces for women. Such work helped travel include comfortable and familiar spaces for women. The early days of steam transit presented new opportunities for women; their writings and actions show that they carefully weighed these options before making choices. Etiquette books and anxious fathers offered advice on how to travel. While some women preferred to travel alone and some wanted the oversight of a chaperone, all women appear to have been well aware of the stakes of that choice and what the selection said to the public about their own independence and respectability. The encounters that white women had with steam transit illustrate the naturalizing process as women entered new and unfamiliar spaces and show how American culture reflected back on steam transit as corporations worked to accommodate these women.

CHAPTER SEVEN

Children

In the fall of 1860, nine-year-old H. D. Barnard sat at a printing press "on a hard stool by a dirty window" and composed a small pamphlet. Despite his young age, when he selected a topic for his pamphlet, he already had enough experience as a traveler that he titled his work *Travels by Land and Water*. He filled the brief account with his experiences traveling across the United States. In 1851, Barnard took his first journey when he was two months old and traveled from his birthplace of Detroit to New York City. He had no memories of this trip: "All I know about it is I was told I made a great noise in the cars." In a later journey, he returned to Detroit. He recalled that his family lost the key to the luggage, that he fell ill, and that he "liked to hear fairy stories." He remembered even more about a later journey from New York to Wisconsin: "Whizzing past houses and villages, now going through a long tunnel hewn in solid rock, then rushing like fire through fields and forests, with our iron horse, we went on until one afternoon I fell asleep, and waking up amid the creaking of breaks, the ringing of bells and various other noises, I found myself at journey's end." Reaching the end of his booklet, Barnard extended the idea of travel metaphorically to his own writing: "I hope, as I said in the Preface, that you are satisfied with this Book to which I bid good bye, and also to the type. I have now got to my journey's end."[1]

Barnard was young when he traveled and young when he wrote the book. Yet even this young man had inculcated the language of travel used by countless other writers. His writing demonstrates how the youngest generation of Americans—those who grew up with steam transit—understood these developments. First, it was possible for a child living in the 1850s to travel enough at a young age that these adventures could admirably fill a self-published pamphlet. Second, these travels were important and meaningful enough to him that he made it the focus of his writing, even though the typesetting and printing were "hard work." Travel was not only physically possible but also personally meaningful. Third, young Barnard had clearly picked up on the common descriptive language used to describe railroads at the time: "whizzing," "rushing like fire," and "iron horse." Fourth, the sensory experience was critical to Barnard's experience on the train. Of all the things that he remembers, the aural encounter was key: the "creaking" of brakes, the bells sounding as the train pulled into the depot, and "various other noises" were all memorable.

This chapter explores the interaction that children had with steam transit. Children themselves have not generally left us a great deal of primary source material from which historians can draw. Barnard's direct account is a rarity. The majority of this chapter, then, focuses on literature that was written for children in the antebellum era. One reason studying this literature is valuable is that such literature "uncovers the values that society hopes to transmit to its children."[2] Much of the children's literature of the antebellum period had a moral purpose, whether that purpose was founded in a moral republicanism or an evangelical Christian didacticism. Examining these works demonstrates how new technologies could be used to these educational ends.

A second reason is that if children's literature communicates information about the values taught to children, then it must necessarily also teach us about antebellum adults—those who wrote the lessons and determined that they were appropriate for children. Adults in the early republic could, of course, remember a time before canals and railroads existed, so for them these transportation improvements were genuinely new. As witnesses to this change, they had to sort out the meaning of these changes. Examining the role of transportation in children's literature shows us what truths antebellum adults wanted to communicate to their children about technology. Through this literature, adults took their accumulated experience with technological change and gave it to their children, but these lessons were not just about technology itself. As we have seen, steam transit had tremendous impacts on culture and allowed people to move rapidly from their homes. The antebellum era was one of dislocation, and in this literature adults wrestled with the question, if transportation could help move people away from what was familiar, were there still essential truths that needed to be communicated to children?[3]

A third reason that examining children's literature during the era of steam transit is fruitful is that steam transit emerged at a specific time in the history of childhood. Modern childhood was "invented" during the early nineteenth century. During this time, the concept that childhood should be "free from labor and devoted to schooling," which was a more "sheltered" view than previously held by American parents, began to emerge.[4] Even if efforts to provide compulsory schooling were incomplete in the antebellum era, there was no doubt that parents—particularly in the urban middle- and upper-class areas and spreading outward—were beginning to adjust their child-rearing techniques. The years of adolescence became of particular importance during the early nineteenth century. Just when this sheltered view of childhood emerged, steamboats and railroads offered independence. Culture demanded attachment to the family, but with improved transportation, children could easily move beyond familial bonds. As transportation possibilities opened up for children, adults had to contend with what these changes meant and how they felt children should be prepared for them. Children were not fully formed adults. They possessed partial independence while still being partially dependent on adults.[5] In this literature, steam transit showed children how they could be more independent and responsible. There is admittedly a bias here toward children who were literate and—in the case of readers or spellers—were attending school. Nevertheless, these themes are present in a wide range of literature, including from the era's most prolific children's authors. The resulting portrait shows how lessons about both steam transit and life were transmitted to the next generation.

Children's Interaction with Steam Transit

There are a few direct accounts that hint at children's engagement with steam. In one example, Elizabeth Speed, who was a student at the Briercliff School in Aurora, New York, in 1823, implored her brother, who lived in Ithaca, to come for a visit. She urged that the trip would not inconvenience him: "You can come in the Steam Boat one day and return the same."[6] In another example from 1848, John Long was about nine years old and living in Buckfield, Maine. He kept a diary and recorded the progress of the railroad. That November, he noted that the "oldest man in the town," at ninety-nine years old, "shoveled the first shovelful of dirt" to start construction of a new railroad. The following year, he noted that the railroad would be completed "as soon as the Company can raise money" to do the necessary work. In 1850, he wrote that a storm prevented the railroad from coming through and delivering the mail. Finally, in an entry for 1852, he complained that engineers were beginning to survey a railroad on his family's land. "It is going through the very best part of our field and will almost destroy its looks and value. But if it were not extended Mr. Smith would take up the iron on this road and we should have none at all."[7] Long thus showed his understanding—perhaps from hearing adults discuss the same issues—of the trade-offs inherent in railroad construction. Twelve-year-old Samuel Thorne wrote in his journal in 1848, "We arrived in Alexandria just one minute too late for the 2 o'clock boat. Therefore we had to wait for the 4 o'clock one."[8] Even as a youth, Thorne began to inculcate the importance of being on time. These examples show us how youth were experiencing and processing their interactions with steam transit, but this type of documentation remains a rarity.

Children who grew up in a world with steamboats and railroads could not know what a world without them looked like. One reason for this is that toy versions were quickly made for children to play with. Toy trains entered the marketplace in the 1830s, immediately after actual trains appeared on the landscape. The most basic versions were simply pushed along the floor by toddlers. Specialized toys arrived shortly thereafter. Joseph Paxson of Bucks County, Pennsylvania, built a miniature coal hopper in 1840. Sawyer's Works in Portland, Maine, built a toy locomotive and tender to commemorate the Atlantic and St. Lawrence Railroad. In 1848, a toy manufacturer in Philadelphia advertised a toy train made in honor of President Zachary Taylor.[9] In 1856, toymaker George Brown made the "first known clockwork train."[10] These toys were "imitative," which "help a child mimic adult life, in preparation for the 'real world.'"[11] As miniature versions of the real thing, they helped acclimate children to a world with machines. This acclimation was useful, because initial exposure to these machines could be frightening. For example, in 1852, Chastina Rix recorded in her diary that her son, no more than two years old, was "very much afraid of the cars."[12] Toys helped children become familiar with machines prior to their first encounter. The machine was brought into the home and made manageable—something the child could easily pick up, examine, and use to create their own imaginative stories.

Other types of evidence demonstrate that parents entrusted their children in using steam transit. Boarding schools, for example, added information about proximity to steam transit to their advertising, suggesting that it would be easy for students to commute home when necessary. In 1838, the Pittsfield Commercial and Classical Boarding School noted that it was "forty-two miles by rail-road, from Hudson, N. Y.; and about equally distant, by the same mode of conveyance, from Springfield, Mass."[13] An advertisement for the Quaboag Seminary in Warren, Massachusetts, featured an image of a train pulling up in front of the school building (fig. 7.1). The advertisement boasted that "the Seminary and Mansion command a fine view of the cars, which, beside affording a convenient means of coming to and passing from the place, are a constant source of pleasurable excitement and interest."[14] Other schools advertised their proximity to transit, including those in New Jersey, Connecticut, New York, Vermont, and New Hampshire.[15] Parents took advantage of the nearness of the railroads to schools. In 1852, Lincoln Clark asked his wife if she could "go down by stage and railroad and back the same day if necessary" in order to check on their daughter who was at school.[16] When another of Clark's daughters went to school, she did so by train and wrote to her sister that she "found when we arrived at the Dêpot that there were eight young ladies besides myself for Miss P's school."[17] For a certain class of parents, then, steam transit allowed easier access to schooling.

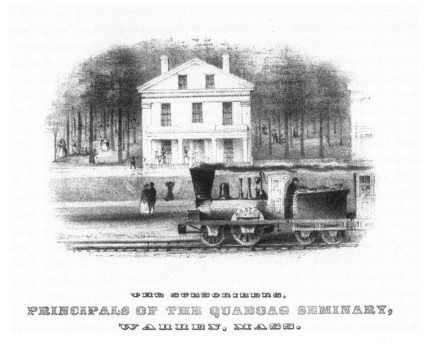

Figure 7.1. An advertisement for Quaboag Seminary in Warren, Massachusetts, from the late 1840s, demonstrates the convenience of the railroad in providing access to the school. Catalog record 151649. *Courtesy, American Antiquarian Society.*

Newspapers also provide some hints about how children interacted with steam transit. In 1859, a newspaper credited a boy for offering an old man his seat on a train. When the old man inquired why the boy had made the offer, he responded, "Because you are old, sir, and I am a boy!" The author was much gratified by this response, and "wanted to seize hold of the little fellow, and press him to my bosom."[18] The boy's willingness to give up the seat was praiseworthy. In a similar vein, *Frederick Douglass' Paper* published a notice about how a young boy had bravely prevented a railroad accident. This boy, Eli Rheem, had witnessed a bridge collapse, and then "remembering that the express passenger train was then about due from York, started off at the top of his speed to endeavor to stop the train, which he knew must be close at hand." When the train drew near, Rheem jumped onto the track and flagged down the engineer, who brought his train to a halt. Rheem's willingness to put his own life in danger made an enormous difference, according to the newspaper, because boys had a bad reputation for disrupting trains. The paper noted that "engineers are so awful cheated by mischievous boys on the route, that they seldom pay attention to them." Therefore, only by placing his life directly in peril did Rheem convince the engineer that he should stop the train. The passengers of the train "liberally rewarded the boy for his presence of mind and daring," and the company's board of directors gave him an additional $100.[19]

Tragically, accidents could even claim children. Children may have grown up with steam transit, but it does not mean that they automatically understood the risks involved and the danger posed by these massive machines. In 1832, a group of boys in Philadelphia were "amusing themselves" with a railroad car and were apparently pushing it down a track. Their horseplay resulted in the death of one of the boys.[20] Likewise, the *Lafayette* (Indiana) *Daily Journal* reported on a funeral for a young boy who had been killed while playing near a train in 1854. The newspaper wrote that the "funeral exercises were solemn and imposing and well calculated to impress upon the minds of children, the danger of playing about the railroad."[21] Children did not know a world other than that which included steam transit, but they still needed to be educated that steam transit could be deadly.

Transportation and Progress

Much of the children's literature of the era emphasized the progress that steam transit represented. This portion of the literature was firmly in step with other booster literature of the age. According to one history of schoolbooks, "the wonders of the machine age are celebrated in glowing terms throughout the century."[22] One such book was a dictionary for youth written in 1832 by John Longgley, who included entries for "Canals" and "Steam." While the entry on canals traced the history of those improvements back to ancient times, the steam entry looked forward in anticipation: "There is no doubt that machines impelled by the force of steam are the most powerful ever formed by the art of man," Longgley informed his young readers, and compared the power of a steam engine equivalent to "the combined action of two hundred good horses," a collection of animals that provided children with a way to gauge the immense power of the railroad but also must have staggered the imagination.[23] Other works for children invoked praise of steam as well. Lyman Cobb's *North American Reader* included De Witt Clinton's "Eulogy on Chancellor Livingston and Robert Fulton." Clinton praised steamboat inventor Fulton as the "Archimedes of his country."[24] Children saw that inventors were singled out for praise.

School books explained how Americans specifically benefited from steam. One 1835 reader included a description of the Hudson Valley to demonstrate to children how nature could be transformed by technology. The valley "as it was" was not an attractive sight, and children were told of its previous, repellant appearance: "Reptiles sported in the stagnant pools, or crawled unharmed over piles of mouldering trees." Even the sound of the valley was frightening: an unproductive "silence reigned," interrupted only by occasional birds' wings and the "howl of beasts of prey." The result of this unfruitful landscape had a palpable effect on humans. The untamed countryside left the Native Americans as "wild as the savage scene, . . . a vagrant over the continent, in constant warfare with his fellow-man." In this racist vision, the author argued that the result of uncultivated land was the debasement of human morality. But the valley soon changed, and transportation technology (canals, bridges, and railroads) was a key part of this transformation. The impact of the transportation improvements could be felt on the morality of the community: "Manners are made benevolent by civilization; and the virtue of the country is the guardian of its peace." The human condition, as exhibited in the author's racial hierarchy, was directly improved as a result of man having "subdued, cultivated and adorned" nature. Transportation improvements were directly linked with improvements to Hudson Valley society at large and the replacement of native people by white people.[25] Comfort was here found in the technology, not provided by nature.

The same reader included another piece titled "The West: A Glance at the Future." The author projected out as to what the future might hold for the western United States. As of the time of the writing, he noted, "As yet no rail-road intersects the western prairie beyond the Arkansas and Mississippi." But this surely would not last long; the author urged the reader to look fifty years into the future: "See yonder steam-car darting across the prairie, having in its train a hundred passengers.... What a change! The splendid steamer now disturbs the waters of the Mackenzie and the Columbia." The effect on humans was noticeable: "Yonder shrivelled Indian is the last of his race; his people are no more—his hunting ground hath yielded to the plough—his wigwam is destroyed—and he stands solitary and alone, the last relic of a mighty race." Progress, brought by trains and steamboats, would destroy the old way of life. This transformation was marked by a change in sound: the region would one day hear "the sound of the hammer of the artisan... and the woodsman's axe shall resound through the forest." This progress would have a clear human cost, presented to the student as tragic but inevitable.[26]

Students also read about how transit connected them to the wider world. One lesson, in an 1848 reader, was titled "The Privileges I Enjoy in My Own Country" and boasted of the material conditions of American life and the non-white labor that made it possible. "In China, men are gathering the tea-leaf for me; in our own soil, they are planting cotton for me," the lesson stated, referencing without specifically naming slavery. These non-white men who provided benefits for white children were joined by machines as well: "At home, powerful steam engines are spinning and weaving and making clothes for me, and propelling vessels and rail cars, to convey me, with speed, from one place to another." The labor of non-white people was equated to that of the machine, and both were in service of the white child reading the book. Transportation brought a host of benefits that the book enumerated: "I have coaches running night and day to carry my correspondence. I have roads, canals, and bridges to bear the coal for my winter fires." The final lesson was gratitude to God for such bounty: "This picture is not exaggerated, but might be extended; such being the goodness of God, that each individual of the civilized millions covering the earth, may have, by very simple means, nearly the same enjoyments, as if he were lord of all."[27] The local wharf or depot that a child might know well was but a small link in a vast system of transportation. The improvements wrought by this transportation seemed endless. Steam transit brought material improvement, to be sure, but there was also a moral lesson for the youth: the "goodness of God" brought the advantages for the children of the United States to benefit from.

Images could also excite children about the possibilities of steam travel. *The Little Keepsake*, published in 1849, included a poem called "The Railroad Ride." The poem went, "Ding dong,/Off we go;/Sing a song,/Hurra ho./Ding dong,/Here we are;/Hold in strong,/Stop the car./Ding dong,/Back we go;/Fly along,/Hurra ho!" The poem emphasized safety ("hold in strong") as well as the speed of the train ("fly along"). The accompanying image also considered safety, showing a train with several passenger cars rushing down a track. The track is in a culvert with high embankments on both sides (towering over the train itself). A dismounted figure restrains a horse from crossing the track. Above the man and horse stands the familiar sign "Look Out for the Locomotive" (fig. 7.2). The image fully conveys the excitement of the oncoming train and the crossing.[28]

Boosters continuously made "predictions of political and moral improvement" throughout the antebellum era.[29] For this class of literature, transportation did not bring with it anxiety or uncertainty. Some of this spirit of improvement was present in the literature for children. The adults who penned these passages—or selected them for children to read—had no doubts

Figure 7.2. The familiar warning for pedestrians to "look out for the locomotive" appears in a children's book. *The Little Keepsake: A Poetic Gift for Children* (New York: Kiggins and Kellogg, 1849). Catalog record 218438. *Courtesy, American Antiquarian Society.*

as to the positive changes transportation would bring. Antebellum boosters worked hard to convince adults that transportation would bring an array of benefits, and these arguments trickled down to literature written for children as well. School literature of the time "assure[d] the American child that his country has done more than any other" in the area of industrial and transport developments.[30] The vast array of machines were working tirelessly to bring the child material gain. The message for children was clear: they were fortunate to live in such a country that had such technological gifts.

Lessons Taught by Transportation

Beyond praising technological advancement, authors used children's literature to teach lessons, both practical and moral. This was in keeping with larger trends in literature at the time. In the wake of the Second Great Awakening, there was an "outpouring" of books intended for young people, with the hopes of preparing them for "life of duty, to both God and society."[31] In addition to manuals with an expressly didactic purpose in mind, there was also a considerable amount of fiction written for children that featured moral instruction. According to historian Daniel T. Rogers, children's fiction expanded considerably after 1830, and this literature was replete with "deliberately commonplace events, a pervasive reasonableness, and insistent moral choices."[32] The types of normal events covered in this fiction included steam transit. A railroad or steamboat ride could be the thrilling centerpiece of a children's story. But more critically, technology could help deliver an important moral lesson. As historian Ronald Zboray has argued, "some Americans turned to novels and stories to help them address the personal challenges of rapid development and the diverse emotional experiences it brought."[33] Literature written for children was no exception. Children's literature also delivered more practical lessons.

One genre that authors used to pass on lessons was the school reader. In 1829, the *United States Reader* published a fictional story about Elizabeth McDermott, who went to Baltimore with her parents. In the story, Elizabeth learned about the railroad, and in so doing, the youth reading the story learned a practical lesson about railroads as well. Elizabeth was no bumpkin. The "wonders" of Baltimore "did not surprise her, as she had been there before, and seen other cities." But there was one thing that was completely new to her: the railroad. She expressed concern to her father: if a road was laid with the type of rails used for fences, surely it "would break the horses' feet, the carriages and our bones, to drive along it." Her father assured her that a railroad was made of different rails. There was no danger of broken bones or feet, because it was "the smoothest of all roads," and was like a "moving house." He promised that the whole family would go to learn about railroads, gave Elizabeth a journal and pencil, and instructed her to take notes. The *Reader* then reproduced her "journal," which reflected her journey on the railroad. She admitted that she did not think she would understand how a railroad worked, but when she saw it, "all became plain enough." Although the railroad was still being constructed, Elizabeth wrote that she could envision the cars as "little moving houses, where you and I could sit and read, as well as we could under the chestnut tree." She concluded with enthusiasm that riding a railroad was "fine sport."[34]

In this story, the fictional Elizabeth learned about railroads and how to be comfortable with them, thus preparing her young readers to have a similar reaction when they encountered a train. Her reaction, comparing railroads several times to houses, emphasized the machine as domestic and familiar, not mechanical and foreign. The fictional father gave Elizabeth and the reader a practical lesson on how to think about railroads and how not to fear them upon first sight.

Other types of literature gave practical advice. In an 1857 novel titled *Whistler; Or, the Manly Boy*, the fourteen-year-old "Whistler" begins his steamboat journey with an important lesson. The bell of the steamer "is pealing forth its last call" while a late passenger arrives by stagecoach, his arm "sawing the air" in "vain" hopes of holding back the ship. But the ship pulls off, and the late passenger is the subject of ridicule: "A general half-suppressed laugh from the crowd on the wharf and the steamer reminds the unhappy straggler that there is something ridiculous, as well as provoking, in being a little too late; and, seeking refuge in the carriage, he is leisurely driven off, to be again laughed at, perchance, when he reaches the home he had lately left in such hot haste." The message to children was clear: to be late was a laughable offense twice over—from the strangers on the boat and the family at home. Children were thus taught the value of being on time. Society scorned the man who was late. If children reading the book were already familiar with the demands of time management, they could laugh at the latecomer along with the characters in the story; if not, they would quietly learn a lesson and thus spare themselves embarrassment in the future.[35]

Whistler learned another important lesson about manners on the trip. Once on board, he heard foreign visitors make some criticisms of the Americans: "'Well, sir,' added the first speaker, 'I've breakfasted with the Turks, I've dined with the Arabs, I've supped with the Chinese, and I've eaten with nearly all the nations of Europe; but, sir, I must say that I never met with such a greedy, scrabbling set of gormandizers as I have found in this country. Why, sir, they seize and devour their food like wild beasts." Whistler's "pride and patriotism were both touched" upon hearing these comments, and "he made up his mind that he would never be guilty of such rudeness, either at a public or a private table," lest he earn a similar rebuke from foreign observers. The author's intent was that his young readers would learn the same lesson and adjust their behavior accordingly.[36]

In addition to social norms like manners and time management, young readers could learn about the actual mechanics of transportation. Sometimes this was done with a simple metaphor. In Jacob Abbott's popular series of Rollo books, young Rollo learns how a steam boiler operates via a comparison to an apple roasting over the fire. Such a lesson brought the power of the steam engine down to the smaller scale that young readers would understand.[37] Other works for children were more detailed, satiating their appetite to learn everything they could about this technological marvel. Elisha Noyce's *The Boy's Book of Industrial Information* was a treasure trove of technically oriented information, describing the operation of pistons, boilers, railroads, canals, locks, and a host of other technologies. The readers of this book could get the technical information they craved.[38]

Not only were technical concepts conveyed, but steam transit permeated basic lessons in terms of spelling and vocabulary. *Comly's Reader* taught the process of building the railroad and then drilled students on the lesson: "SPELL rail-road; sought; which; digging; valleys; levelled; foundation; either; pieces; wheel; carriages; iron; fastened; reaching; instead; without; axles; friction; define it." In this case, the lesson not only communicated specific information about the railroad; it also carried lessons in spelling and vocabulary.[39] In another instance, a spelling book's section on three-syllable words included the word *disaster*. After learning the word, students then used it in a sentence: "The blowing up of the Fulton at New-York was a terrible disaster."[40] Even tragic steamboat accidents had the possibility to teach a lesson.

Other primers also turned to steam transit for their lessons on vocabulary and counting. The *National Pictorial Primer* used railroads to teach about multisyllable words and how steam engines operated. A picture of a railroad with tender, passenger car (with people sitting on top of the roof exposed to the elements and engine smoke), and a waving flag accompanied the story (with words broken up into syllables):

How fast the cars move! That lit-tle car in front is a steam en-gine on wheels; there is a fire in it, and a large i-ron boil-er, with wa-ter. When it gets hot it turns in-to steam; then the guide rings his bell. Ding, dong, bell! Ding, dong, bell! Ting! Then the long train moves off on the i-ron road. Puff! puff! puff! It is al-most out of sight. It goes twice as fast as a horse can run, and does not get tired. Five or six hun-dred peo-ple can go on a jour-ney at one time on a railroad.

In this lesson, students would not only learn about railroads but expand their vocabularies. Elsewhere, a picture of a Mississippi steamboat and the closing page of the book used steamers and locomotives to teach numbers: "Two Steamers which in rivers keep; / Three Coaches—how the whips do Crack! / Four Locomotives on the track."[41]

Such lessons on timeliness, manners, mechanics, spelling, vocabulary, and numeracy were no doubt important. But by far the most frequent way in which transportation was invoked in children's literature was for lessons of morality. Steamboats and railroads were recruited by authors who doubtless saw advantages in using modern and captivating technologies to make their broader points. In 1832, Jacob Abbott wrote *Young Christian*, which was later excerpted in other readers. For example, B. D. Emerson's *Second-Class Reader* (1841) featured an extended story from *Young Christian* about a steamboat. According to the Bible, wrote Abbott, our stay on this earth is "a state of probation, that is, of *trial and discipline*." To help the young reader understand the meaning of this trial, Abbott invoked "some familiar examples, drawn from the actual business of life."

Abbott's "familiar example" was the steamboat. He noted that when a new steamboat is built, it "must be *proved* before put to service." Abbott enumerated the things that could possibly go wrong with a ship (poor-quality materials, poor construction) and noted that the engineer carefully inspected every portion of the vast new machine in order to ensure its proper working before launch. This inspection was not "idle curiosity," Abbott warned, but was "to discover and remedy every little imperfection, and to remove every obstacle which prevents more entire success." Abbott detailed the tasks of the engineer: "He scrutinizes the action of every level and the friction of every joint; here he oils a bearing, there he tightens a nut; one part of the machinery has too much play, and he confines it—another too much friction, and he loosens it; now he stops the engine, now reverses her motion, and again sends the boat forward in her course." Abbott's engineer continued "for many days, or even weeks," inspecting every valve and gauge. After the engineer was satisfied to the quality of the work, only then could "one long procession of happy groups" entrust their lives to the boat.

Thanks to the engineer's diligence, the steamboat could now proceed, embodying both power and restraint: "Loaded with life, and herself the very symbol of life and power, she seems something ethereal—unreal, which, ere we look again, will have vanished away. And though she has within her bosom a furnace glowing with furious fires, and a reservoir of death—the elements of most dreadful ruin and conflagration—of destruction the most complete, and agony the most unutterable; and though her strength is equal to the united energy of two thousand men, she restrains it all." By the close of Abbott's story, the meaning was clear to the student. The young reader of the story, like the steamboat described therein, has "within you susceptibilities and powers, of which you have little present conception, energies, which are hereafter to operate in producing fulness of enjoyment or horrors of suffering, of which you now but little conceive. You are now on *trial*." Just as the engineer had to closely inspect the steamboat before launching and taking lives into its care, so too should the reader "look within, to examine the complicated movements of your heart, to detect what is wrong, to modify what needs change and rectify every irregular motion." Like a mechanic readying a steamship, American youth were to inspect and perfect their hearts. "You are *on trial—on probation* now," Abbott admonished his readers. "You will enter upon *active service* in another world." For Abbott, the steam engine presented an opportunity to reaffirm moral stories about how one should live one's life. Abbott was able to recruit steam transit to give his Christian message.[42]

A poem about a steamboat wreck communicated the moral lesson that men and women should not get too proud, but remember that life could end at any moment. The poem opened with a brief introduction explaining the circumstances of the wreck: "The Steamer Atlantic was wrecked, in a storm, on Long Island Sound, in Nov., 1846. As soon as the boat struck, its bell commenced tolling, probably from the action of the wind upon it, and continued to toll slowly and mournfully, as long as any portion of the wreck was to be seen." The poem then recounted the various people for whom the bell tolled: "the master bold," "the man of God," "the lover, lost," "the absent sire," and so on. This poem used the accident to describe the range of humanity present on the ship and the terrible consequences of the accident. The lesson was about the danger of pride: "Tell how o'er proudest joys / May swift destruction sweep."[43] Steam transit represented a high point of human achievement, but danger was never far away. Children reading this work were to heed the lesson that death could strike at any moment and they should be prepared for the afterlife.

Transportation also taught lessons about gratitude. An 1848 reader includes a story about a group of boys boisterously playing during a break from school: they play with kites, balls, and hoops or simply try to see who can run or jump the farthest. One of the boys stands apart from the group because of an injury: "When he was very young, he was so much injured in one of his feet by a rail car, that he could only walk with the help of a cane." Yet this boy's unfortunate condition did not prevent him from having a good time: "He enjoys the sight, and appears to take as much pleasure in the sports of the rest, as if he could participate with them in their exercises." The injured boy brought a lesson to the reader: that it is a "mark of selfishness, when a person can not have enjoyment in seeing others happy. This lame boy has the whole school for his friends, because they not only feel for his afflictions, but know that he does not complain, and that he takes delight in their company." The review questions at the end of the chapter drove home this message: "5. What is said of the lame boy with a cane? 6. Ought we not to rejoice to see others happy?" For the student readers of this story, the moral lesson was that personal misfortune—here caused by a train—should not prevent children from being grateful for the lives they had. Railroads were certainly dangerous, but here the lesson was that crippling caused by modern technology should not lead children to abandon a lesson taught that predated the railroad: that of the error of selfishness.[44]

Children's literature also warned children about transportation's dangers. An example is a book of stories published in 1855 that included a vignette called "The Depot." An accompanying image shows adults and children awaiting an oncoming train, with luggage by their side (fig. 7.3). The text assures us that there "are many very handsome depots in the United States furnished with every thing that will afford comfort for travellers. The cars too are sometimes very beautiful." But the railroad was not an unmixed blessing. The authors immediately turned to the danger of the cars: "Accidents very often happen on rail-roads, and lives are often lost by the carelessness of those having charge of the locomotive." This danger was communicated not just through a measurement of speed but by describing the disorientation of the senses: "They go very fast; indeed so fast, that you cannot see the houses, or trees along the road." The authors warned children that beneath the image of the depot and beautiful cars lay potential danger from accidents.[45]

Figure 7.3. Children and adults await the oncoming train in this illustration from a children's book. *The Skating Party, and Other Stories* (New York: Leavitt & Allen, 1855). Catalog record 221130. *Courtesy, American Antiquarian Society.*

Authors also emphasized that changes in technology and the modernization of society did not alter the role of religion or morality in life. After noting that the growth of transportation had a beneficial effect for other animals, since "the oppression of stage horses, as formerly driven, is avoided or lessened," a reader noted that the proper expansion of railroads "might be a benefit to the human family, for which gratitude should arise to Him who teacheth man knowledge and discretion." The extension of technological benefit into the religious realm even encompassed the spelling terms from this lesson, which included not only passengers, carriage, and locomotive but also oppression, gratitude, and discretion. The construction of railroads and the benefits they brought with them were seen as another reason to praise God, not as a reason to reject religion for secular life.[46]

Transportation could lead to adventure for fictional characters in children's literature, and those adventures brought moral lessons. Readers of *What Norman Saw in the West* were treated to an extensive itinerary of everything that the Transportation Revolution allowed Norman to experience on a lengthy trip from New York through the Great Lakes region and into Minnesota. The book is truly about what Norman sees and hears: the landscape speeding by his train, missionaries in Illinois returned from China, his numerous relatives on farms throughout the Midwest, the mighty Mississippi River, the many tales of Native Americans, Niagara Falls, fireworks on the fourth of July, harvest time, Toronto, Chicago, St. Paul, and a host of other sights and sounds. Railroad and steamboat travel made this wide range of experiences possible. For the young readers of this book, railroads seem to open up boundless opportunities for exploration. But religion was never far away. As the trip came to a close, "It was with deep gratitude to God that Norman and his mother retired to rest that evening. They were thankful that his kind providence had watched over them in their journey of more than three thousand miles, and had brought them home again, to find those whom they loved well and happy."[47] The message, closing with God's providence, was that religion still had a role to play, even in a world where technology made extraordinary journeys possible.

Other examples can be drawn from novels. Several novels of Jacob Abbott—the author of the piece above who compared preparing a steamboat to preparing for life—made use of technology for major plot points. Indeed, as historian Katherine Pandora has noted, Abbott's "welcoming tone" and "unintimidating entry points into learning" made it easier for his readers to understand the technological changes happening around them.[48] In his novel *Aunt Margaret*, our hero is a young boy named John True. The True family lives in New York City, and the decision is made that John and his younger sister Lucy should visit their Aunt Margaret in Franklin, Massachusetts.

The discussion leading up to this railroad trip immediately becomes a lengthy meditation on fear, responsibility, and a host of other issues. John's father felt that this railroad journey had the opportunity to play an important part in the children's upbringing: "I think that the great danger which we have to fear in respect to our children, is, that they will grow up inefficient and helpless, on account of being waited upon and taken care of so much." By contrast, Mr. True noted, he and his wife "were thrown in a great measure upon our own resources and responsibility from our earliest years." Mr. True was worried that the city life was causing his children to be weak. "From their infancy we put them under the care of a nursery maid," he said, "who has nothing to do all the day long but to wait upon them in every conceivable way." He was troubled that "under pretense of saving them from getting hurt," children would not "ever acquir[e] any substantial experience in respect to the laws of life." Mr. True's fears comported exactly with the transition in child-rearing at the time, as middle-class urban parents began to shelter their children. Mr. True worried that this may have gone too far.[49]

Mr. True then revealed his particular concerns about John. "I want him to be made a *man*," Mr. True said, "a full grown, strong, and energetic man." But the city lifestyle of the True family made this a challenge. Speaking to his wife, Mr. True continued, "But it is extremely difficult, living as we do, to avoid his growing up weak, helpless, and dependent. All that we can do is to watch for occasions to throw responsibility upon him, as, indeed, you have always done. And now we have here an excellent occasion. Let him undertake the charge of himself and Lucy to Franklin." Here lay the crux of the matter. Dependency was long feared by American men as something to be avoided: the antithesis of "republican manhood." Mr. True worried that his child's upbringing would lead him to be too dependent on others. This trip provided the perfect opportunity to demonstrate that John could act independently. Even an accident could be beneficial, his father reasoned, because it would give John the opportunity to exercise "judgment and discretion," which were "the very best lessons he can have." The railroad was a school for John: a lesson on how to manage one's self out in the world and care for the woman (his younger sister) entrusted to his care. In so doing, the fictional Trues adopted another realistic practice of their era: encouraging the children not to be fearful of new situations.[50] As historian Patricia Crain has argued, Abbott's novels placed children in a "knowable, secular world, whose materials and practices a child can learn to master, and whose regulations and mores he—and quite often she—can comprehend and navigate." In these stories, the act of navigating a railroad trip shows the young reader that the world is knowable and can be mastered.[51]

Abbott gave John True opportunities to use his judgment, and the first time John had to do so, he faltered. Upon arriving at the station where the transfer was required, John found himself "fully taken up with his eagerness to get to the other train" and completely forgot "his mother's injunction to do nothing in a hurry." As a result, John and Lucy boarded the wrong train, and now they were "going on at the rate of twenty miles an hour out of their way." The conductor noticed their mistake when checking their tickets, and John and Lucy got off at the next station in hopes of catching a return train. The stationmaster there suggested an alternative: hiring a wagon to take them across to the correct rail line, where they could catch the afternoon train. John thus had to decide how to spend his money: take the wagon over to the road or take the train back and stay overnight. He is forced to make a quick judgment, and then made his decision: "He placed great confidence in the station-master's recommendation. He thought that he, occupying, as he did, an important post, was to be regarded as a responsible man, who would not give any advice to travelers coming to his station unless it was really good advice. So he at length concluded to go." After the wagon journey, they boarded the correct train, arrived safely, and were welcomed by their aunt. In Abbott's hands, the railroad functioned as an effective school for John True. He was faced with challenges, was forced to respond, and ultimately selected the correct course. Railroads gave youth independence, and in so doing gave them opportunity to exercise the values that predated the technology.[52]

In another novel by Abbott, railroads again gave a boy an opportunity to demonstrate judgment and independence. In *The Three Gold Dollars*, recently orphaned Robin Green began walking to New York City to seek a place to live. During his walk, for the first time in his life, he heard a railroad whistle. Despite the fact that he had never seen a railroad before, he clearly had heard about them and knew their significance: "'Ah!' said he, 'there are the cars. I suppose if I were in the cars I should get to New York in two hours.'" Robin mounted a snowbank to view the train and pronounced it to be "Amazing! . . . How like lightning they go!" Robin's unfamiliarity with the train became an opportunity for the author to explain the railroad to the reader. Abbott provided a brief engineering lesson, describing the sleepers and how the rails were fastened to them. Robin pushed against the rail with his foot and was very impressed that it did not give way.

Robin was soon to get an even closer introduction to the train. While walking, he met a flagman—the man responsible for signaling train engineers when danger lay ahead. After Robin and the flagman met, a train stopped before a broken sleeper and the flagman had to race back in order to signal the following train to stop. But he fell and injured himself, and he instructed Robin to take his place. Robin did, flagging down the train and preventing a collision. The conductor of the second train suspected Robin of "roguery," but when the slowed train came upon the injured flagman and his wife, the conductor made amends and offered to take Robin to New York City, his intended destination. As had John True, Robin Green demonstrated the value of quick action and decision-making. Abbott intended his young readers to accept the lesson of responsible action that Robin demonstrated, and the railroad provided the perfect venue for illustrating the lesson, with its potential for danger averted by Robin's quick action.[53]

The exploits of orphans were not limited to young boys such as Robin. In the novel *A Summer with the Little Grays*, seven-year-old orphan Kate traveled from relative to relative, since no one wanted her. On board a railroad she met the Gray family. The parents of the Gray family were a bit astonished that Kate was traveling alone, but she was clearly experienced with traveling and delighted passengers with her stories. Meeting Kate had a transformative effect on the Gray parents. Mr. Gray informed his wife, "I have learned a lesson about children's travelling alone. I believe I should not be afraid to send [their daughter] Emmy off to-morrow for a journey of a hundred miles. Just think of this little thing [Kate] utterly alone, and of the friends she has made in one short trip!"[54] Orphans such as Robin and Kate demonstrated their ingenuity in part through their encounters with modern transportation. As middle- and upper-class parents became more concerned with their children and sheltering them from the world, these books showed that independence and self-reliance were still useful attributes. Indeed, the exploits of Robin and Kate demonstrated to young readers that transportation was not something that only adults could navigate.

In 1852, the first volume of the *Youth's Casket; an Illustrated Magazine for Children* appeared from the press of Beadle and Brother in Buffalo, New York. Among the stories printed was the tale "A Trip to the Falls," in which young Emily and Clara travel with their cousin and mother to Niagara Falls. The trip involves a steamboat ride, which excites the young girls. They return home via railroad. The author of the story writes that when the train approached, Emily and Clara "heard a bell, and the peculiar noise a locomotive makes, when the steam is up—you all know how it sounds better than I can spell it."[55] Here, the author counted on the fact that his young audience would be familiar with the sound of the train. The author brought the reader into his or her confidence and let their own knowledge of steam transit feed their imagination rather than trying to dictate the sounds that Emily and Clara heard.

The author's gamble that the reader would be acquainted with steam transit was a wise one. By 1852, steam transit had been a feature of American life for decades, and the audience for the *Youth's Casket* would have grown up with steam transit. Direct primary source evidence from youth about their encounters with steam is fleeting, but the examples that we do have, such as Barnard's self-published book, underline that steam transit had an impact on youth as well as on adults. Moreover, authors used transportation to teach a wide range of lessons to the youth who were their readership. Some of these lessons were about the technology itself or used technology to teach other skills such as spelling. But far more often, there was a deeper meaning. The steamboat could be a metaphor for the well-examined life, or a railroad trip could give a young person the opportunity to demonstrate an ability to act independently in the world. For the authors considered here, these were powerful lessons to be communicated to a new generation. In the hands of these authors, trains and steamboats were tools to teach children crucial moral lessons. These adult authors lived through turbulent times but still felt there were truths about religion and behavior worth communicating to children. Transportation played a role in creating this turbulence but also became the methodology for teaching the next generation. For antebellum youth, steam transit would always be a natural part of their surroundings, and the literature considered here demonstrates how earlier generations passed on their accumulated experience and wisdom.

Conclusion

While the political and economic impact of antebellum steam transit has been well examined by historians, steam's impact was not limited to the economy. The Transportation Revolution permeated American culture and society, and American culture, in turn, put its imprint on steam transit as well. There is a larger story to the Transportation Revolution beyond freight and passengers.

The interplay between steam transportation and culture happened on multiple levels. Steam transit was conceived and operated in communities large and small across the United States. Corporations had to convince residents that steam transportation was worth pursuing. As construction began, landholders and corporations fought for what each believed was fair. For many Americans, the first trip on a steamboat or railroad was revelatory, worthy of comment in a letter to friends or privately in a journal. As the networks grew, steam transportation commodified travel, transforming the process from one that required individual travelers to make their own arrangements to one that was purchased in the form of a ticket. Passengers began to create their own norms to regulate behavior on board and accepted the level of risk that steam transit brought with it for deadly accidents.

Steam's impact quickly moved far from the wharves and tracks and permeated many dimensions of American culture. The pressure of steam and the speed of a railroad were ready metaphors in both public writing and private correspondence. Humorists quickly picked up on the astonishment created by the opportunities offered by steam, and over time they poked fun at those who were unfamiliar with steam transit and its conventions. Composers took steam transit as a theme, producing musical works to be enjoyed by a crowd at an orchestra concert or in the privacy of one's own home on a piano and in some instances mimicking the sound of transportation itself. Images of steam transit flourished, from the large scale of a moving multi-hour panorama to the small scale of currency carried in one's pocket. A burgeoning market for travel guidebooks allowed potential travelers to preview their routes and armchair travelers to imagine their own voyages. Religious thinkers incorporated steam transit into their spiritual messages, with many Protestant ministers seeing the development of steam transit as a sign of God's blessing on the United States. While there was always the temptation that steam transit could lead people to concentrate too much on worldly things, on the whole Protestant ministers saw it as yet another way to expand the reach of God's word. As Americans turned to their churches for comfort after deadly steam accidents, preachers responded by seeing in those accidents a warning that believers should prepare themselves for death at any moment.

Foreign commentators noted the open design of American steamboats and trains, which allowed for significant social mixing. This meant that many different types of passengers could interact while on board: people of different genders, races, and classes could travel together. Black Americans in the North and in the South had distinctly different experiences. For enslaved people, the transportation networks they built under duress became a potential route for escape. While transportation for white people was commodified, enslaved people seeking escape created their own routes through inventiveness and skill, taking advantage of the networks as best they could. In the North, depots and wharves became flashpoints as crowds gathered to protest Black people taken or returned to the South against their will. And Black Americans fought against discrimination in the North, demanding to be treated as equals in the public sphere. For their part, white women faced a choice when considering transportation. Steam transit held out the possibility of travel alone, but that required navigating complex public space. Travel with a chaperone could be safer, as the accompanying man would deal with all interlocutors. Women selected both options throughout the antebellum era and clearly understood the implications of their decisions. Finally, children grew up in a world with steam transit, meaning that it always seemed perfectly normal. For the adults who raised them, incorporating steam transit into literature for children gave authors the opportunity to teach valuable practical and moral lessons.

The wide range of issues considered in this book would continue into the postbellum era, as historians of that era have ably demonstrated. But the antebellum interplay of steam transit and culture considered here are evidence for the deep-seated roots of these postbellum phenomena. From the beginning of steam transit, railroads and steamboats moved out into the broader culture. Americans considered their meaning and then pressed steam transit into service literally and metaphorically for a wide range of uses. It was an experience that was broadly shared across the country. Yet an experience in common did not mean that it was unifying: white women in New England, enslaved men in the South, free Black people along the Ohio River—all had different experiences with these technologies. Of course, other technologies throughout American history would also inspire widespread adoption and debate about their cultural impact; as I write these words in August 2023, Americans are still grappling with the cultural changes that the internet has wrought on our lives. Just as H. D. Barnard and other children grew up with steam transit, so have today's children grown up with technological opportunities that seemed unimaginable to their ancestors. The question that antebellum ministers posed—to what use will this technology be put, good or ill—still resonates as modern Americans confront the technologies that define our age as surely as steam defined the antebellum era.

The experience of steam transit was not unifying, but if the lack of unity was a cost, the blossoming of reactions to steam chronicled in these pages was surely a benefit. The people who developed American steam transit perhaps never envisioned that music would be composed in their honor or that authors of fiction would use steamboats and trains as story settings. They surely did not intend that railroads and steamboats would help aid enslaved people escape from slavery. Could they foresee that women would be emboldened to travel on their own, that depots intended for the storage of goods would instead be pressed into service for religious ceremonies, that children would push toy trains across the floors of their homes, or that George Gordon Byron DeWolfe of New Hampshire would call himself the "Steam Machine Poet" and hand out his poems in railroad depots?[1] Steam power fired the imaginations of all types of Americans, who put their own stamp on steam transit's meaning. Americans Black and white, northern and southern, urban and rural, men and women, and young and old looked at steam transit and saw not just a means of transit but possibilities for freedom; they were inspired to write poetry and music and created a grand cacophony of cultural output that eventually reached the doorstep of Sallie McNeill. The creators of this cultural mosaic had different motivations and desires, but the sum total of their work was to create an environment that naturalized steam transit for those who had never seen it before.

NOTES

Abbreviations

AAS	American Antiquarian Society, Worcester, Massachusetts
AAS-B	American Antiquarian Society broadside collection
BARC	Boston and Albany Railroad Collection, Baker Library, Harvard Business School, Cambridge, Massachusetts
BL	Baker Library, Harvard Business School, Cambridge, Massachusetts
CA	*Colored American*
FDP	*Frederick Douglass' Paper*
GLB	*Godey's Lady's Book*
HL	The Huntington Library, San Marino, California
HSP	Historical Society of Pennsylvania, Philadelphia
KJV	King James Version Bible
LCP	Library Company of Philadelphia, Pennsylvania
LoC	Library of Congress, Washington, DC
LR	*Ladies' Repository*
MHS	Massachusetts Historical Society, Boston
NE	*National Era*
NS	*North Star*
SCL	South Caroliniana Library, University of South Carolina, Columbia
VHS	Virginia Historical Society, Richmond
WHS	Wisconsin Historical Society, Madison

Introduction

1. Sallie McNeill, *The Uncompromising Diary of Sallie McNeill, 1858–1867*, ed. Ginny McNeill Raska and Mary Lynne Gassaway Hill (College Station: Texas A&M University Press, 2009), 3, 10, 21, 33, 54, quotation on 57. Here, McNeill used the common term *cars* to refer to a train. For an exploration of this term, see Aaron Marrs, "What's in a Name? 'Railroad' vs. 'the Cars,'" https://aaronwmarrs.com/blog/2015/02/whats-in-name-railroad-vs-cars.html. Primary sources from the antebellum era feature several different spellings of railroad (chiefly "rail-road" and "rail road"). When quoting from primary sources, I have always reproduced the spelling found in the original document. In my own writing and in the names of corporations, I have standardized on "railroad."

2. For an overview of the economic or political impact of the Transportation Revolution, see George Rogers Taylor, *The Transportation Revolution, 1815–1860* (New York: Rinehart, 1951); Albert Fishlow, "Internal Transportation in the Nineteenth and Early Twentieth Centuries," in *The Cambridge Economic History of the United States*, vol. 2, *The Long Nineteenth Century*, ed. Stanley L. Engerman and Robert E. Gallman (New York: Cambridge University Press, 2000), 543–642; and John Lauritz Larson, *Internal Improvement: National Public Works and the Promise of Popular Government in the Early United States* (Chapel Hill: University of North Carolina Press, 2001). For a general discussion of technology and American culture, see Leo Marx, *The Machine in the Garden: Technology and the Pastoral Ideal in America*, rev. ed. (New York: Oxford University Press, 2000); Bruce Kuklick, "Myth and Symbol in American Studies," in *Locating American Studies: The Evolution of a Discipline*, ed. Lucy Maddox (Baltimore: Johns Hopkins University Press, 1999), 71–86; Jeffrey L. Meikle, "Leo Marx's *The Machine in the Garden*," *Technology and Culture* 44 (2003), 147–59; and John F. Kasson, *Civilizing the Machine: Technology and Republican Values in America, 1776–1900*, rev. ed. (New York: Hill and Wang, 1999). Another critical work, although written about the European experience, is Wolfgang Schivelbusch, *The Railway Journey*, trans. Anselm Hollo (New York: Urizen, 1979), see esp. chap. 3. One counterexample to the trend of treating transportation separately is John Lauritz Larson, *The Market Revolution in America: Liberty, Ambition, and the Eclipse of the Common Good* (New York: Cambridge University Press, 2010), which weaves a discussion of transportation through much of the narrative. Daniel Walker Howe has proposed using *communications revolution* for the era. Daniel W. Howe, *What Hath God Wrought: The Transformation of America, 1815–1848* (New York: Oxford University Press, 2007), 5. For a general overview of major themes in railroad historiography, see Albert Churella, "Company, State, and Region: Three Approaches to Railroad History," *Enterprise and Society* 7 (2006), 581–91.

3. Will Mackintosh, "'Ticketed Through': The Commodification of Travel in the Nineteenth Century," *Journal of the Early Republic* 32 (2012), 64.

4. Edward E. Baptist, *The Half Has Never Been Told: Slavery and the Making of American Capitalism* (New York: Basic Books, 2014), 351, 366–69; Sven Beckert, *Empire of Cotton: A Global History* (New York: Knopf, 2014), 191, 218, 240; Calvin Schermerhorn, *The Business of Slavery and the Rise of American Capitalism, 1815–1860* (New Haven, CT: Yale University Press, 2015), 176–79, 192–98, and chap. 7; and Walter Johnson, *River of Dark Dreams: Slavery and Empire in the Cotton Kingdom* (Cambridge, MA: Belknap Press of Harvard University Press, 2013), esp. chaps. 3–5 and pp. 6, 256–57, 293–96, 319–20. For an overview of the scholarly literature and public debate over slavery and capitalism, see Matthew Pratt Guterl, "Slavery and Capitalism: A Review Essay," *Journal of Southern History* 81 (2015), 405–20; Christopher Morris, "With 'the Economics-of-Slavery Culture Wars,' It's Déjà Vu All Over Again," *Journal of the Civil War Era* 10 (2020), 524–57; and John Lauritz Larson, "The Genie and the Troll: Capitalism in the Early American Republic," *Journal of the Early Republic* 42 (2022), 589–622.

5. David E. Nye, *America as Second Creation: Technology and Narratives of New Beginnings* (Cambridge, MA: MIT Press, 2003), 2.

6. A. K. Sandoval-Strausz, *Hotel: An American History* (New Haven, CT: Yale University Press, 2007); Richard H. Gassan, *The Birth of American Tourism: New York,*

the Hudson Valley, and American Culture, 1790–1830 (Amherst: University of Massachusetts Press, 2008); Elizabeth Stordeur Pryor, *Colored Travelers: Mobility and the Fight for Citizenship before the Civil War* (Chapel Hill: University of North Carolina Press, 2016); Will Mackintosh, *Selling the Sights: The Invention of the Tourist in American Culture* (New York: New York University Press, 2019); and David Schley, *Steam City: Railroads, Urban Space, and Corporate Capitalism in Nineteenth-Century Baltimore* (Chicago: University of Chicago Press, 2020). For a work that thinks more broadly about the impact of railroads on American society, ranging from its earliest years to the present, see Julia H. Lee, *The Racial Railroad* (New York: New York University Press, 2022).

7. For postbellum treatments of steam transit, see Amy G. Richter, *Home on the Rails: Women, the Railroad, and the Rise of Public Domesticity* (Chapel Hill: University of North Carolina Press, 2005); Barbara Young Welke, *Recasting American Liberty: Gender, Race, Law, and the Railroad Revolution, 1865–1920* (New York: Cambridge University Press, 2001); Alex Ruuska, "Ghost Dancing and the Iron Horse: Surviving through Tradition and Technology," *Technology and Culture* 52 (2011), 574–97; Thomas G. Andrews, "'Made by Toile'? Tourism, Labor, and the Construction of the Colorado Landscape," *Journal of American History* 92 (2005), 837–63; Eric Arnesen, *Brotherhoods of Color: Black Railroad Workers and the Struggle for Equality* (Cambridge, MA: Harvard University Press, 2001); Greg Umbach, "Learning to Shop in Zion: The Consumer Revolution in Great Basin Mormon Culture, 1847–1910," *Journal of Social History* 38 (2004), 29–61; Laura Elaine Milsk, "Meet Me at the Station: The Culture and Aesthetics of Chicago's Railroad Terminals, 1871–1930" (Ph.D. diss., Loyola University, 2003); and Scott R. Nelson, *Iron Confederacies: Southern Railways, Klan Violence, and Reconstruction* (Chapel Hill: University of North Carolina Press, 1999). For the importance of social history, see Maury Klein, *Unfinished Business: The Railroad in American Life* (Hanover, NH: University Press of New England, 1994), 166–86. For examples of social history, see Eugene Alvarez, *Travel on Southern Antebellum Railroads, 1828–1860* (University: University of Alabama Press, 1974); James A. Ward, *Railroads and the Character of America, 1820–1887* (Knoxville: University of Tennessee Press, 1986); Carol Sheriff, *The Artificial River: The Erie Canal and the Paradox of Progress, 1817–1862* (New York: Hill and Wang, 1996); John R. Stilgoe, *Train Time: Railroads and the Imminent Reshaping of the United States Landscape* (Charlottesville: University of Virginia Press, 2007); Aaron W. Marrs, *Railroads in the Old South: Pursuing Progress in a Slave Society* (Baltimore: Johns Hopkins University Press, 2009); and Craig Miner, *A Most Magnificent Machine: America Adopts the Railroad, 1825–1862* (Lawrence: University Press of Kansas, 2010).

8. For the growing dominance of railroads, see Kasson, *Civilizing the Machine*, 172. For steamboats' economic performance, see Fishlow, "Internal Transportation," 565, table 13.5, and 568, table 13.6; and Robert Gudmestad, *Steamboats and the Rise of the Cotton Kingdom* (Baton Rouge: Louisiana State University Press, 2011), chap. 8. For different technologies working together, see Richard W. Bulliet, "Determinism and Pre-Industrial Technology," in *Does Technology Drive History? The Dilemma of Technological Determinism*, ed. Merritt Roe Smith and Leo Marx (Cambridge, MA: MIT Press, 1994), 205; and Michael Cotte, "Railways and Culture: An Introduction,"

in *Eisenbahn/Kultur = Railway/Culture*, ed. Günter Dinhobl (Innsbruck: Studien Verlag, 2004), 45.

9. Marrs, *Railroads in the Old South*, 5, table 2.

Chapter 1 • Community Relations

1. See, for example, Ronald Kline and Trevor Pinch, "Users as Agents of Technological Change: The Social Construction of the Automobile in the Rural United States," *Technology and Culture* 37 (1996), 763–95. Unless otherwise noted, any emphasis noted in quotations is in the original, and all quotations have been reproduced as found in the original (including spelling errors).

2. Joseph Caldwell, *The Numbers of Carlton, Addressed to the People of North Carolina on Central Railroad through the State* (New York: G. Long, 1828), 56–58 (quotation on 56). The quoted essay was first published October 22, 1827, and a series of essays were then collated in a single publication cited here.

3. David Schley, *Steam City: Railroads, Urban Space, and Corporate Capitalism in Nineteenth-Century Baltimore* (Chicago: University of Chicago Press, 2020), 25. For the changing conceptions of "steam" that were emerging around the same time, see David Philip Miller, "A New Perspective on the Natural Philosophy of Steams and Its Relation to the Steam Engine," *Technology and Culture* 61 (2020), 1129–48.

4. *Proceedings of a Convention Holden at Dover, N.H. September 29, 1835, on the Subject of a Rail Road from Portland to Boston* (N.p.: n.p., 1835), 1, AAS.

5. "Morgan's Newly Invented Rail Road Carriage," 1829, AAS-B.

6. Charles Endicott to Francis Endicott, April 20, 1823, folder 5, box 1, Endicott Family Papers, AAS.

7. Harriott Pinckney Horry, June 5, 1815, 1815 Journal, in *The Papers of Eliza Lucas Pinckney and Harriott Pinckney Horry Digital Edition*, ed. Constance Schulz (Charlottesville: University of Virginia Press, Rotunda, 2011), http://rotunda.upress.virginia.edu/PinckneyHorry/ELP1072.

8. Nathan Foster Diary, AAS.

9. August 19, 1835, Caroline B. Poole Diary, LoC.

10. July 14, 1841, Evan Randolph I travel diary, "Boston Trip," folder 11, box 14, series 2c, Carson-Randolph Family Papers (3004), HSP.

11. Entry for October 18, 1848, vol. 2, box 1, Elizabeth Steele Wright Papers, LoC.

12. Craig Miner, *A Most Magnificent Machine: America Adopts the Railroad, 1825–1862* (Lawrence: University Press of Kansas, 2010), xiv.

13. For an overview of this genre, see Leon Jackson, "We Won't Leave Until We Get Some: Reading the Newsboy's New Year's Address," *Common-place* 8 (January 2008), http://commonplace.online/article/we-wont-leave-until-we-get-some/.

14. "The Carrier's address to the patrons of the *Patriot & chronicle*," 1828, AAS-B.

15. "Carrier's address to the patrons of the *Doyleston Democrat*," 1829, AAS-B.

16. David Trowbridge Brigham, "Public sale of 75 building lots," 1836, AAS-B.

17. *A Detailed and Correct Account of the Grand Civic Procession, in the City of Baltimore, on the Fourth of July, 1828; in Honor of the Day, and in Commemoration of the Commencement of the Baltimore and Ohio Rail-Road* (Baltimore: Thomas Murphy, 1828), 14–33, quotations on 21, AAS.

18. *Detailed and Correct Account of the Grand Civic Procession*, 23.

19. Schley, *Steam City*, 1, quotation on 38–39.

20. William A. Link, *Roots of Secession: Slavery and Politics in Antebellum Virginia* (Chapel Hill: University of North Carolina Press, 2003), 34–35.

21. Charles H. Hildreth, "Railroads out of Pensacola, 1833–1883," *Florida Historical Quarterly* 37 (1959), 407.

22. Addie Lou Brooks, "The Building of the Trunk Line Railroads in West Tennessee, 1852–1861," *Tennessee Historical Quarterly* 1 (1942), 112.

23. John Harkins, "Memphis," entry in the *Tennessee Encyclopedia*, https://tennesseeencyclopedia.net/entries/memphis/.

24. Quoted in Martin Bruegel, *Farm, Shop, Landing: The Rise of a Market Society in the Hudson Valley, 1780–1860* (Durham, NC: Duke University Press, 2002), 87.

25. Matthew W. Klingle, "Spaces of Consumption in Environmental History," *History and Theory* 42 (2003), 94–95.

26. David E. Nye, *America as Second Creation: Technology and Narratives of New Beginnings* (Cambridge, MA: MIT Press, 2003), 176.

27. Solomon W. Roberts, "Reminiscences of the First Railroad over the Allegheny Mountain," *Bulletin* [of the Railway and Locomotive Historical Society] 44 (1937), 13.

28. *Evidence Showing the Manner in Which Locomotive Engines Are Used upon Rail-Roads: And the Danger and Inexpediency of Permitting Rival Companies Using Them on the Same Road* (Boston: Centinel & Gazette Press, 1838), 9, 19, AAS.

29. *Eighth Annual Report of the Directors of the Western Rail-Road Corporation to the Stockholders, Presented February 8, 1843, Comprising a Copy of the Seventh Report to the Legislature* (Boston: Dutton and Wentworth's Print, 1843), 30, 31, AAS.

30. *Sixth Annual Report of the Little Miami Rail Road Co., December, 1848* (Cincinnati: E. Shepard, 1848), 6, AAS.

31. William Hasell Wilson, *The Columbia-Philadelphia Railroad and Its Successor* (1896; repr., York, PA: American Canal and Transportation Center, 1985), 11–12.

32. David Schley, "A Natural History of the Early American Railroad," *Early American Studies* 13 (2015), 466.

33. Alfred Rix and Chastina W. Rix, *New England to Gold Rush California: The Journal of Alfred and Chastina W. Rix, 1849–1854*, ed. Lynn A. Bonfield (Norman, OK: Arthur H. Clark, 2011), 253.

34. Lincoln Clark to Julia Annah Clark, June 6, 1852, CL 295, Lincoln Clark Papers, HL.

35. Lee A. Craig, Raymond B. Palmquist, and Thomas Weiss, "Transportation Improvements and Land Values in the Antebellum United States: A Hedonic Approach," *Journal of Real Estate Finance and Economics* 16 (1998), 183, 185. Canals could also increase values and economic activity. See Whitney R. Cross, *The Burned-Over District: The Social and Intellectual History of Enthusiastic Religion in Western New York, 1800–1850* (Ithaca, NY: Cornell University Press, 1950), 59.

36. Chad Coffman and Mary Eschelbach Gregson, "Railroad Development and Land Value," *Journal of Real Estate Finance and Economics* 16 (1998), 198.

37. Link, *Roots of Secession*, 34.

38. Tom Downey, *Planting a Capitalist South: Masters, Merchants, and Manufacturers in the Southern Interior, 1790–1860* (Baton Rouge: Louisiana State University Press, 2006), 96–97.

39. Frederick B. Gates, "The Impact of the Western & Atlantic Railroad on the Development of the Georgia Upcountry, 1840–1860," *Georgia Historical Quarterly* 91 (2007), 183–84.

40. Walter Gwynn Turpin to "My dear wife," May 31, 1857, section 1, Walter Gwynn Turpin Papers, VHS.

41. Quoted in S. Max Edelson, *Plantation Enterprise in Colonial South Carolina* (Cambridge, MA: Harvard University Press, 2006), 206.

42. Quoted in *Connecticut Farmer's Gazette and Horticultural Repository* (May 1844), 196.

43. *Farmer's Monthly Visitor* (January 1845), 7.

44. Entry for June 11, 1844, Letterbook (1842–1853), Stoughton Branch Railroad Collection, BL.

45. Charles Russell to Stephen Kensick[?], November 10, 1847, folder F-7-32, carton 6, Baker Vertical Files, BL.

46. Letter from "Yr. obt. srvt." to "Dear Sir," September 29, 1849, folder labeled "Correspondence, etc., relating to 1849–52," Virginia and Tennessee Railroad Materials, Norfolk and Western Railroad Historical Collection, VHS.

47. Michael J. Connolly, *Capitalism, Politics and Railroads in Jacksonian New England* (Columbia: University of Missouri Press, 2003), 26.

48. Petition of David McElwain to the Directors of the Western Railroad, September 18, 1840, folder labeled "Western Railroad, Clerks File, 1841," case 1, BARC.

49. Petition, October 2, 1840, folder labeled "Western Railroad, Clerks File, 1841," case 1, BARC.

50. Letter from David McElwain to "Messrs," April 17, 1842, folder labeled "Western Railroad, Clerks File, 1841," case 1, BARC.

51. Manuscript report to the directors' meeting of May 18, 1842, folder labeled "Western Railroad, Clerks File, 1842," case 1, BARC.

52. J. C. Gray to J. Foster, June 22, 1846, box 2, case 848, Richard Henry Dana Papers, AAS.

53. Paul N. Edwards et al., "AHR Conversation: Historical Perspectives on the Circulation of Information," *American Historical Review* 116 (2011), 1309.

54. J. C. Gray to S. M. Felton (July 22, 1846), J. C. Gray to S. M. Felton (August 2, 1847), and Nathaniel Wyeth to John Gray (January 12, 1848), all in box 2, case 848, Richard Henry Dana Papers, AAS.

55. *Fourth Annual Report of the Directors of the Pennsylvania Railroad Co. to the Stockholders, December 31, 1850* (Philadelphia: Crissy & Markley, 1851), 17.

56. *Report of the Committee of Investigation Appointed by the Stockholders of the Boston and Maine Railroad, at a Meeting Held at Exeter, N.H., May 28, 1849* (Boston: Eastburn's Press, 1849), 20.

57. Asa Sheldon, *Yankee Drover: Being the Unpretending Life of Asa Sheldon, Farmer, Trader, and Working Man, 1788–1870* (1862; repr., Hanover, NH: University Press of New England, 1988), 122–23.

58. *Memorial of the President and Directors of the Philadelphia, Easton & Water-Gap Railroad Company, to the Councils of the City of Philadelphia* (Philadelphia: McLaughlin Brothers, 1853), 4.

59. A. K. Sandoval-Strausz, *Hotel: An American History* (New Haven, CT: Yale University Press, 2007), 55.

60. Petition dated [ca. June] 1837 to the Western Railroad, folder labeled "Western Railroad, Clerks File, 1837," case 1, BARC.

61. Petition dated October 22, 1839, folder labeled "Western Railroad, Clerks File, 1841," case 1, BARC.

62. Thomas B. Wales to George Bliss, May 3, 1836, Letterbook of T. B. Wales, vol. 145, Western Railroad, BARC.

63. George Stark to Artemus Parker, Charles Wheeler, and others, July 21, 1853, vol. 111, Nashua and Lowell Railroad Collection, BL.

64. *Fifteenth Annual Report of the President and Directors of the Baltimore and Susquehanna Rail-Road Company* (Baltimore: James Lucas and E. K. Deaver, 1843), 7, AAS.

65. Subcommittee's report on the western terminal, contained in the minutes of the Executive Committee in Springfield, June 8, 1841, Albany and West Stockbridge Railroad Company, vol. 2, BARC.

66. Portsmouth and Roanoke Railroad Company, "Regulations to be observed by those sending or receiving produce or other articles," 1836, AAS-B.

67. "Rates of transportation on the Petersburg, Greensville and Roanoke, and Raleigh and Gaston Railroads," 1840, AAS-B. For additional examples of rate-setting, see William Parker, "Boston & Worcester Rail Road. Freight tariff for merchandize generally . . . ," 1845; Connecticut River Railroad Company, "Connecticut River rail road . . . 1845," 1845; Norwich and Worcester Railroad Company, "Regular steam boat line between Norwich and New York," 1845; and "1845–6. Freight tariff. Schenectady and Troy and Troy and Greenbush Railroads," 1845, all in AAS-B.

68. Charles Eastman to Bardwell, Smith, and Company, June 17, 1828, Charles Eastman Letterbook, AAS.

69. George Stark to Thomas D. Smith, June 21, 1853, vol. 111, Nashua and Lowell Railroad Collection, BL.

70. Schley, *Steam City*, 127.

71. *Report of the Majority of the Special Committee on the Subject of Removing the Rails of the Harlem Rail Road Company, South of Fourteenth Street* (New York: n.p., 1841), 512. For contemporary reporting by the Black press, see *CA* (March 6, 1841).

72. Martin J. Hershock, *The Paradox of Progress: Economic Change, Individual Enterprise, and Political Culture in Michigan, 1837–1878* (Athens: Ohio University Press, 2003), 40–48 (quotations on 40–41, 43).

73. James Thomas and others to the President and Board of Directors of the Richmond, Fredericksburg and Potomac Railroad, April 1, 1859, BR 193-1e, Brock Collection, HL.

74. Newburyport (MA), "By-laws for the Town of Newburyport," 1843, AAS-B.

75. Chester (PA), "Ordinance!" 1851, AAS-B.

76. Charlotte (NC), "Ordinances of the Town of Charlotte," 1859, AAS-B.

77. Directors' meeting of March 28, 1836, minutebook of the board of the directors and stockholders (1835–1837), Boston and Worcester Railroad, vol. 2, BARC.

78. G. Twichell to W. A. Bryant, January 17, 1850, G. Twichell Letterbook, Boston and Worcester Railroad, vol. 119, BARC.

79. *Whately Herald* (November 8, 1850).

80. *East Whately Herald* (January 1851).

81. See Wiebe E. Bijker, *Of Bicycles, Bakelites, and Bulbs: Toward a Theory of Sociotechnical Change* (Cambridge, MA: MIT Press, 1995), for an introduction to this literature.

82. Directors' meeting of November 17, 1836, minutebook of the directors and stockholders (1835–1837), Boston and Worcester Railroad, vol. 2, BARC.

83. *LR* (July 1849), 197.

84. *Seventeenth Annual Report of the Boston and Worcester Railroad Corporation* (Boston: Dutton and Wentworth, 1849), 10, AAS.

85. Tom Moore Craig, ed., *Upcountry South Carolina Goes to War: Letters of the Anderson, Brockman, and Moore Families, 1853–1865* (Columbia: University of South Carolina Press, 2009), 21.

86. *Fourth Annual Report of the Directors of the Northern Rail-Road Corporation, to the Stockholders, May, 1849* (Boston: Press of Crocker and Brewster, 1849), 15.

87. *Twenty-Second Annual Report of the Directors of the Boston & Worcester Railroad Corporation, for the Year Ending November 30, 1851* (Boston: David Clapp, 1852), 10, AAS.

88. Meeting minutes of September 1 and 8, 1834, minutebook of the directors and stockholders (1831–1835), Boston and Worcester Railroad, vol. 1, BARC.

89. Massachusetts law quoted in the *Act of Incorporation and By-Laws of the Eastern Rail Road Company, Incorporated April 14, 1836* (Salem, MA: Palfray and Chapman, 1836), 21. For the legal implications of these aural signals, see James W. Ely, Jr., *Railroads and American Law* (Lawrence: University Press of Kansas, 2001), 126.

90. Mark M. Smith, *Listening to Nineteenth-Century America* (Chapel Hill: University of North Carolina Press, 2001), 111; and Deborah Lubken, "Joyful Ringing, Solemn Tolling: Methods and Meanings of Early American Tower Bells," *William and Mary Quarterly*, 3d ser., 69 (2012), 832–33.

91. *Regulations for the Government of the Transportation Department of the Western Rail Road Corporation* (Springfield, MA: Merriam, Wood & Co., 1840), 7–8.

92. G. Twichell to C. C. Andrews, June 5, 1851, G. Twichell Letterbook, Boston and Worcester Railroad, vol. 119, BARC.

93. G. Twichell to E. Washburn, February 5, 1850, G. Twichell Letterbook, Boston and Worcester Railroad, vol. 119, BARC.

94. William Appleton, *Selections from the Diaries of William Appleton, 1786–1862* (Boston: Merrymount Press, 1922), 185.

95. "Citizens of Massachusetts! Read the following, and consider . . . ," 1845, AAS-B.

96. John W. Bear, *The Life and Travels of John W. Bear, "The Buckeye Blacksmith." Written by Himself* (Baltimore: Binswanger and Company, 1873), 110.

97. Quoted in Robert Gudmestad, *Steamboats and the Rise of the Cotton Kingdom* (Baton Rouge: Louisiana State University Press, 2011), 21.

98. Whig Party (New London County, CT) Committee, "New London County Whig Meeting," 1840, AAS-B.

99. *FDP* (October 15, 1852); Angela F. Murphy, *The Jerry Rescue: The Fugitive Slave Law, Northern Rights, and the American Sectional Crisis* (New York: Oxford University Press, 2016), 149.

100. George M. Dexter to S. M. Felton, October 2, 1850, and memorandum dated October 10, 1850, both in folder F-7-23, carton 6, Baker Vertical Files, BL. For Lind's tour and popularity, see Daniel Cavicchi, *Listening and Longing: Music Lovers in the Age of Barnum* (Middletown, CT: Wesleyan University Press, 2011), 14–19, 33–36, 104–5, 149–52.

101. *NE* (October 17, 1850).

102. *Jenny Lind Comic Almanac, 1851* (New York: Elton & Co., [1850]), AAS.

103. *Common School Journal* (June 15, 1840), 185–86.

104. George Templeton Strong, *Diary*, vol. 1, *Young Man in New York, 1835–1849*, ed. Allan Nevins and Milton Halsey Thomas (New York: Macmillan, 1952), 108.

105. *Knickerbocker* (June 1848), 557.

106. Horace Greeley, "The Age We Live In," *Nineteenth Century* (January 1848), 50.

107. Levi Beardsley, *Reminiscences: Personal and other Incidents; Early Settlement of Otsego County; Notices and Anecdotes of Public Men; Judicial, Legal and Legislative Matters; Field Sports; Dissertations and Discussions* (New York: Charles Vinten, 1852), 218.

108. Caroline Healey Dall, *Selected Journals of Caroline Healey Dall*, vol. 1, *1838–1855*, ed. Helen R. Deese (Charlottesville: University of Virginia Press, 2006), 369.

109. "New Year's Address of the Carriers of the *Newburyport Herald*," 1854, AAS-B.

110. D. J. Barber Journal, p. 8, HM 68483, HL.

Chapter 2 • Travel

1. Issac Hinckley, "Newport via Providence," 1848, AAS-B.

2. Will Mackintosh, *Selling the Sights: The Invention of the Tourist in American Culture* (New York: New York University Press, 2019), 64.

3. William Williams to Dorothy Ashley, March 22, 1793, CL 540, Lincoln Clark Papers, HL.

4. Lincoln Clark to Julia Annah Clark, April 18, 1837, CL 107, Lincoln Clark Papers, HL.

5. John H. White, Jr., *Wet Britches and Muddy Boots: A History of Travel in Victorian America* (Bloomington: Indiana University Press, 2013), 46; A. K. Sandoval-Strausz, *Hotel: An American History* (New Haven, CT: Yale University Press, 2007), 15–20. Sandoval-Strausz notes that colonial taverns were "architecturally indistinguishable" from homes because "the vast majority were simply homes with liquor licenses and makeshift bars." Sandoval-Strausz, *Hotel*, 144.

6. Entries for February 7, 9, 15, and 16, 1849, Elizabeth Steele Wright Papers, LoC.

7. Thomas Bull to John Noble, August 8, 1814, NBL 106, Noble Family Papers, HL.

8. June 24, 1825, Alexander Bliss Journal, AAS.

9. Edmund Kirby to Eliza Brown Kirby, April 3, 1828, EK 9, Kirby Papers, HL.

10. October 16, 1828, Augustus Pleasants Webb Journal, HM 29058, HL.

11. Mrs. S. Hopkins to Eliza Bund, December 4, 1812, case 8, box 11, Simon Gratz autograph collection (0205A), HSP.

12. Elizabeth Dowling Taylor, *A Slave in the White House: Paul Jennings and the Madisons* (New York: Palgrave Macmillan, 2012), 61.

13. R. B. Mason to George Mason, August 16, 1821, R. B. Mason Papers, AAS.

14. Henry Waller to William Waller, August 14, 1829, box 1(1827–1829), Waller Papers, HL.

15. Craig Miner, *A Most Magnificent Machine: America Adopts the Railroad, 1825–1862* (Lawrence: University Press of Kansas, 2010), 89.

16. Martha Parker to "My dear Father," June 3, 1834, Leonard M. Parker Papers, AAS.

17. Jack Larkin and Caroline Sloat, eds., *A Place in My Chronicle: A New Edition of the Diary of Christopher Columbus Baldwin, 1829–1835* (Worcester, MA: American Antiquarian Society, 2010), 210. I thank Katheryn Viens for bringing this to my attention.

18. Beverly Scafidel, "The Letters of William Elliott" (Ph.D. diss., University of South Carolina, 1978), 322.

19. Sidney George Fisher, *A Philadelphia Perspective: The Diary of Sidney George Fisher Covering the Years 1834–1871*, ed. Nicholas B. Wainwright (Philadelphia: Historical Society of Pennsylvania, 1967), 51, 54.

20. Alexis Litvine argues against an overreliance on contemporary writers who saw the compression of space as negative. Indeed, the writers considered here do not seem to have viewed spatial compression negatively. Alexis D. Litvine, "The Annihilation of Space: A Bad (Historical) Concept," *Historical Journal* 65 (2022), 893–94.

21. Francis Lieber, *Letters to a Gentleman in Germany: Comprising Sketches of the Manners, Society, and National Peculiarities of the United States* (Philadelphia: Carey, Lea & Blanchard, 1835), 43.

22. January 2, 1842, Richard W. Thompson Papers, LoC.

23. George W. Smith, *Incidents of Travel, from the Pencil Notes of the Author* (Indianapolis: Indianapolis Journal Company, 1855), 78, HL.

24. Scafidel, "Letters of William Elliott," 917.

25. September 27, 1858, Journal of William P. Floyd, HM 19334, HL.

26. "Sketches by the Way," *LR* (September 1843), 275.

27. Henry Waller to Sarah Bell Langhorne Waller, November 27, 1852, box 5(1852), Waller Papers, HL.

28. George Templeton Strong, *Diary*, vol. 2, *The Turbulent Fifties, 1850–1859*, ed. Allan Nevins and Milton Halsey Thomas (New York: Macmillan, 1952), 17.

29. Philadelphia, Wilmington, and Baltimore Rail Road Company, "Philadelphia, August 23, 1837. Sir, I take the liberty . . . ," 1837, AAS-B.

30. February 8, 1841, Meeting of the Board of Directors, carton 4, vol. 2, Norwich and Worcester Railroad, Penn Central Collection, BL.

31. People's Line Steamboats, "People's Line. For New York," 1841, AAS-B.

32. I. H. Tupper, "Evening line of boats . . . ," 1848, AAS-B.

33. Letter from Daniel Helm, November 20, 1847, Daniel Helm Papers, AAS.

34. William Parker, "Boston and Worcester Railroad," 1843, AAS-B.

35. G. Twichell, "First and second class ticket tariff," 1850, AAS-B.

36. Lincoln Clark to Julia Annah Clark, March 11, 1847, CL 197, Clark Papers, HL.

37. Henry B. Tatham, pp. 3–4, Journey to the Ruins of Aztalan, Wisconsin Territory, HM 55636, HL.

38. November 25, 1837, William Banister Diary, AAS.

39. John Kingman, *Letters, Written by John Kingman, while on a Tour to Illinois and Wisconsin, in the Summer of 1838* (Hingham, MA: Jedidiah Farmer, 1842), 14, HL.

40. *Horn's Railroad Gazette* (June 9, 1849), AAS.

41. W. Van Olinda, *Report of Unclaimed and Missing Baggage. No. 3: Chicago, April 15, 1854* (Chicago: Democratic Press Office, 1854), 3, 4, 6, 8, AAS.

42. Henry Waller to Sarah Bell Langhorne Waller (May 28, 1840) and Henry Waller to Sarah Bell Langhorne Waller (June 4, 1840), both in box 3(1840), Waller Papers, HL.

43. "A Trip down the Ohio River," *CA* (September 28, 1839).

44. August 31, 1850, Moses Drury Hoge Diary, VHS.

45. *Rules and Regulations to Be Observed by the Men in the Employment of the Mich. Central Railroad Company. Detroit, June, 1850* (Detroit: Duncklee, Wales, & Co., 1850), 11, 12, AAS.

46. *Finn's Comic Almanac, or United States Calendar, for 1835* (Boston: Marsh, Capen & Lyon, n.d.), n.p.

47. Harriott Pinckney Horry, July 24, 1815, 1815 Journal, in *The Papers of Eliza Lucas Pinckney and Harriott Pinckney Horry Digital Edition*, ed. Constance Schulz (Charlottesville: University of Virginia Press, Rotunda, 2011), http://rotunda.upress.virginia.edu/PinckneyHorry/ELP1119.

48. Julia Wright Sublette, "The Letters of Anna Calhoun Clemson, 1833–1873" (Ph.D. diss., Florida State University, 1993), 103.

49. Eugene Alvarez, *Travel on Southern Antebellum Railroads, 1828–1860* (University: University of Alabama Press, 1974), 54.

50. Directors' meeting of March 10, 1835, minutebook of the directors and stockholders (1835–1837), Boston and Worcester Railroad, vol. 2, BARC.

51. Robert Gudmestad, *Steamboats and the Rise of the Cotton Kingdom* (Baton Rouge: Louisiana State University Press, 2011), 54.

52. Matthew Carey, "Some Notices of Kentucky," 1828, AAS-B.

53. Gudmestad, *Steamboats*, 67.

54. Elisabeth Koren, *The Diary of Elisabeth Koren*, ed. David T. Nelson (Northfield, MN: Norwegian-American Historical Association, 1955), 73, 74.

55. Henry Lunettes, *The American Gentleman's Guide to Politeness and Fashion; or, Familiar Letters to his Nephews, Containing Rules of Etiquette, Directions for the Formation of Character, etc., etc., Illustrated by Sketches Drawn from Life, of the Men and Manners of Our Times* (New York: Derby & Jackson, 1857), 108; more examples follow on 401–11, AAS.

56. *NS* (September 15, 1848).

57. November 26, 1828, Augustus Pleasants Webb Journal, HM 29058, HL.

58. *NE* (July 15, 1858).

59. August 1, 1851, journal 2, Caroline Barrett White Journals, AAS.

60. Wolfgang Schivelbusch, *The Railway Journey*, trans. Anselm Hollo (New York: Urizen, 1979), 66.

61. Gillian Silverman, *Bodies and Books: Reading and the Fantasy of Communion in Nineteenth-Century America* (Philadelphia: University of Pennsylvania Press, 2012), 27.

62. Ronald J. Zboray, *A Fictive People: Antebellum Economic Development and the American Reading Public* (New York: Oxford University Press, 1993), 13.

63. Kevin J. Hayes, "Railway Reading," *Proceedings of the American Antiquarian Society* 106 (1997), 324–25.

64. *Rules and Regulations for the Management of the Philadelphia, Wilmington and Baltimore, and the New Castle and Frenchtown Railroads* (Philadelphia: James Bryson, 1854), 24, AAS.

65. *NE* (May 27, 1852).

66. *NE* (May 31, 1855).

67. *NE* (June 26, 1856).

68. *NE* (February 18, 1858).

69. Hayes, "Railway Reading," 306.

70. Emerson Bennett, *Intriguing for a Princess. An Adventure with Mexican Banditti* (Philadelphia: J. W. Bradley, 1859); *Punch's Pocket-Book of Fun* (New York: D. Appleton, n.d.), both AAS.

71. Hayes, "Railway Reading," 311.

72. Lincoln Clark to Julia Annah Clark, December 10, 1852, CL 328, Clark Papers, HL.

73. Anonymous Travel Diary Transcript, p. 2, WHS (call number SC 894).

74. Michael O'Brien, ed., *An Evening when Alone: Four Journals of Single Women in the South, 1827–67* (Charlottesville: University Press of Virginia, 1993), 158.

75. Carolina Seabury, *The Diary of Carolina Seabury, 1854–1863*, ed. Suzanne L. Bunkers (Madison: University of Wisconsin Press, 1991), 25.

76. Gudmestad, *Steamboats*, 70.

77. April 20, 1843, Dexter Russel Wright Papers, LoC.

78. Philip N. Racine, *Gentlemen Merchants: A Charleston Family's Odyssey, 1828–1870* (Knoxville: University of Tennessee Press, 2008), 78.

79. *NE* (September 28, 1848). For another account of straw polls on trains, see *NS* (September 22, 1848).

80. For an exploration of the discussion of slavery on transatlantic steam voyages, see W. Caleb McDaniel, "Saltwater Anti-Slavery: American Abolitionists on the Atlantic Ocean in the Age of Steam," *Atlantic Studies* 8 (2011), 141–63.

81. *CA* (April 8 and August 5, 1837).

82. *CA* (July 28, 1838; September 7, 1839; and September 19, 1840).

83. *NS* (September 15, 1848).

84. *NE* (March 15, 1849).

85. *FDP* (September 2, 1853).

86. Daniel W. Howe, *What Hath God Wrought: The Transformation of America, 1815–1848* (New York: Oxford University Press, 2007), 429 (quotation), 512–14.

87. For a similar mixing in hotels, see Sandoval-Strausz, *Hotel*, 261.

88. For an overview of major issues in sensory history, see Mark M. Smith, *A Sensory History Manifesto* (University Park: Pennsylvania State University Press, 2021).

89. Sophie du Pont, *Sophie du Pont: A Young Lady in America, Sketches, Diaries & Letters, 1823–1833*, ed. Betty-Bright P. Lowe and Jacqueline Hinsley (New York: Harry N. Abrams, 1987), 144.

90. April 26 [*sic*], 1855, folder 2, David Auguste Burr Papers, AAS. This entry is dated Tuesday, April 26 but is placed between April 23 and 25, so it was presumably April 24.

91. August 21, 1849, journal 1, Caroline Barrett White Journals, AAS.

92. Charlotte Hixon, "Trip to Aiken," March 24, 1849, SCL.

93. Marion Lansing, *Mary Lyon through Her Letters* (Boston: Books, Inc., 1937), 118.

94. August 30 and 31, 1823, Joseph Dowding Bass Eaton Journal, MHS.

95. Anna M. Kauffman to Rudolph and Jane Kauffman, August 26, 1850, HM 77143, Phillips Papers, HL.

96. Henry B. Tatham, p. 31, Journey to the Ruins of Aztalan, Wisconsin Territory, HM 55636, HL.

97. *New Hampshire Gazette and Republican Union* (August 15, 1848).

98. Joseph H. Bragdon, *Seaboard Towns; Or, Traveller's Guide Book from Boston to Portland: Containing a Description of the Cities, Towns and Villages, Scenery and Objects of Interest along the Route of the Eastern Railroad and Its Branches, and the Portland, Saco and Portsmouth Railroad. Including Historical Sketches, Legends, &c.* (Newburyport, MA: Moulton & Clark, 1857), 111, AAS.

99. Scafidel, "Letters of William Elliott," 322.

100. *Richmond Whig* (January 27, 1857).

101. Samuel Chamberlain, *My Confession: Recollections of a Rogue*, ed. William H. Goetzmann (Austin: Texas State Historical Association, 1996), 29.

102. Quoted in Lisa Denmark, "'But What a Pretty Thing a Rail Road Is!—We Must Have One of Our Own—Positively!': Savannah's Unique Rhetoric of Internal Improvements," *Journal of the Georgia Association of Historians* 28 (2009), 15.

103. *A Descriptive Guide to the Eastern Rail-Road, from Boston to Portland; with Brief Notices of the Towns through Which It Passes* (N.p.: n.p., 1851), 32, AAS. For another example of taking meals in different cities, see David Schley, *Steam City: Railroads, Urban Space, and Corporate Capitalism in Nineteenth-Century Baltimore* (Chicago: University of Chicago Press, 2020), 30.

104. Mary Telfair, *Mary Telfair to Mary Few: Selected Letters, 1802–1844*, ed. Betty Wood (Athens: University of Georgia Press, 2007), 174.

105. James D. Davidson, "A Journey through the South in 1836: Diary of James D. Davidson," ed. Herbert A. Kellar, *Journal of Southern History* 1 (1935), 353–54.

106. Lincoln Clark to Julia Annah Clark, June 1, 1843, CL 150, Clark Papers, HL.

107. Edmund Kirby to Eliza Brown Kirby, June 21, 1846, EK 76, Kirby Papers, HL.

108. J. Alexis Shriver, "A Maryland Tour in 1844: Diary of Isaac van Bibber," *Maryland Historical Magazine* 39 (1944), 241.

109. Joseph J. Mersman, *The Whiskey Merchant's Diary: An Urban Life in the Emerging Midwest*, ed. Linda A. Fisher (Athens: Ohio University Press, 2007), 153.

110. July 17, 1856, folder 1, box 1, William Chauncy Langdon Papers, LoC.

111. Henry Waller to Sarah Bell Langhorne Waller, March 4, 1856, box 7(1856), Waller Papers, HL.

112. *Tenth Annual Report of the Board of Directors to the Stockholders of the Memphis and Charleston Railroad Company, July 1, 1860* (Memphis: Wm. M. Hutton and Co., Book and Job Printers, 1860), 49.

113. Mollie Dorsey Sanford, *Mollie: The Journal of Mollie Dorsey Sanford in Nebraska and Colorado Territories, 1857–1866*, ed. Donald Danker (Lincoln: University of Nebraska Press, 1959), 5.

114. William S. Dutt to parents, April 5, 1860, William S. Dutt Correspondence, HL.

115. Schivelbusch, *Railway Journey*, 133; James Taylor, "Business in Pictures: Representations of Railway Enterprise in the Satirical Press in Britain, 1845–1870," *Past and Present* 189 (2005), 130. For a comprehensive history of accidents, see Robert B. Shaw, *A History of Railroad Accidents, Safety Precautions, and Operating Practices* (N.p.: Vail-Ballou, 1978).

116. Drew Gilpin Faust, *This Republic of Suffering: Death and American Civil War* (New York: Knopf, 2008), 6 (quotation), 9–11; Gudmestad, *Steamboats*, 110.

117. Mark Aldrich, *Death Rode the Rails: American Railroad Accidents and Safety, 1828–1965* (Baltimore: Johns Hopkins University Press, 2006), 6, 38 (quotation).

118. Christian Gelzer, "The Quest for Speed: An American Virtue, 1825–1930" (Ph.D. diss., Auburn University, 1998), 208.

119. *Dinsmore's Road and Steam Navigation Guide* (New York: Dinsmore & Co., 1859), 4.

120. Ian R. Bartky, "Running on Time," *Railroad History* 159 (1988), 19.

121. Bartky, "Running on Time," 21 (quotation), 28; Benjamin Sidney Michael Schwantes, *The Train and the Telegraph: A Revisionist History* (Baltimore: Johns Hopkins University Press, 2019), 27; Ian R. Bartky, *Selling the True Time: Nineteenth-Century Timekeeping in America* (Stanford, CA: Stanford University Press, 2000), 94.

122. Gilbert H. Barnes and Dwight L. Dumond, eds., *Letters of Theodore Dwight Weld, Angelina Grimké Weld, and Sarah Grimké, 1822–1844*, vol. 2 (1934; repr., New York: Da Capo, 1970), 665.

123. Charles H. Titus, *Into the Old Northwest: Journeys with Charles H. Titus, 1841–1846*, ed. George P. Clark (East Lansing: Michigan State University Press, 1994), 40.

124. Lincoln Clark to Catherine Lincoln Clark, April 30, 1851, CL 241, Clark Papers, HL.

125. John Williams Gunnison to Alice Mary Gunnison, August 18, 1841, HM 46985, Gunnison Family Papers, HL.

126. Equitable Life Assurance Society of the United States, "Is Your Life Insured?," 1860, AAS-B. "Gasconade Bridge" is presumably a reference to a railroad accident on November 1, 1855, when a railroad bridge collapsed on the Gasconade River in Missouri, killing forty-three people. Walter Johnson, *The Broken Heart of America: St. Louis and the Violent History of the United States* (New York: Basic Books, 2020), 118.

127. Robert Habersham to Barnard Habersham, December 5, 1840, folder 11, box 1, Barnard Habersham Papers, Hargrett Library, University of Georgia, Athens, Georgia.

128. Quoted in Simon Cordery, *The Iron Road in the Prairie State: The Story of Illinois Railroading* (Bloomington: Indiana University Press, 2016), 23.

129. Gudmestad, *Steamboats*, 99 (quotation), 101–5.

130. Ann Archbold, *A Book for the Married and Single, the Grave and the Gay: And Especially Designed for Steamboat Passengers* (East Plainfield, OH: N. A. Baker, 1850), 76–77, HL.

131. "Tribute to Mrs. Anne Hill," *GLB* (December 1852), 576.

132. Andrew Leary O'Brien, *The Journal of Andrew Leary O'Brien* (Athens: University of Georgia Press, 1946), 29–30.

133. Lincoln Clark to Julia Annah Clark, January 14, 1849, CL 226, Clark Papers, HL.

134. September 23, 1849, journal 1, Caroline Barrett White Journals, AAS.

135. *Tricks and Traps of New York, Number 1* (New York: Dinsmore & Co., 1857), 36.
136. *Tricks and Traps of New York 1*, 37–38, 40.
137. *Tricks and Traps of New York, Number 2* (New York: Dinsmore & Co., 1858), 40, 41.
138. "Remarkable & Shocking Deaths, &c.," call number 23843, HL.
139. "Carrier's address to the subscribers of the *Baltimore American*," 1859, AAS-B.
140. Bryan F. Le Beau, *Currier and Ives: America Imagined* (Washington, DC: Smithsonian Institution Press, 2001), 12, 13.
141. "Awful Conflagration of the Steam Boat *Lexington*," E. M. Fowble, *Two Centuries of Prints in America: 1608–1880, A Selective Catalogue of the Winterthur Museum Collection* (Charlottesville: University Press of Virginia, 1987), plate 320.
142. John Collins, "Accident on the Camden and Amboy," folder 7, box 9, American Antiquarian Society Drawings Collection, AAS.
143. April 5, 1851, Lurana Y. White Chase Diary, AAS.
144. Bragdon, *Seaboard Towns*, 136.
145. *NE* (August 5, 1858).
146. A Traveller, *A Letter to the Hon. Daniel Webster, on the Causes of the Destruction of the Steamer Lexington: As Discovered in the Testimony before the Coroner's Jury in New York* (Boston: Charles C. Little and James Brown, 1840), 43–44.
147. Albert Fishlow, "Internal Transportation in the Nineteenth and Early Twentieth Centuries," in *The Cambridge Economic History of the United States*, vol. 2, *The Long Nineteenth Century*, ed. Stanley L. Engerman and Robert E. Gallman (New York: Cambridge University Press, 2000), 580, table 13.9.
148. Faust, *This Republic of Suffering*, 30; Gudmestad, *Steamboats*, 112.

Chapter 3 • The Arts

1. David E. Nye, *America as Second Creation: Technology and Narratives of New Beginnings* (Cambridge, MA: MIT Press, 2003), 2.
2. Thomas Shreve, "The Young Lawyer," *Cincinnati Mirror and Western Gazette of Literature and Science* (February 1835), 143.
3. *NE* (June 8, 1848).
4. *NE* (July 26, 1849).
5. *Crockett Awl-Man-Axe for 1839* (Philadelphia: Turner & Fisher, n.d.), AAS.
6. *Racine* (Wisconsin) *Advocate* (May 25, 1853).
7. *A Descriptive Guide to the Eastern Rail-Road, from Boston to Portland; with Brief Notices of the Towns through Which It Passes* (N.p.: n.p., 1851), 29, AAS.
8. *GLB* (November 1858).
9. Ann D. Gordon, ed., *The Selected Papers of Elizabeth Cady Stanton and Susan B. Anthony*, vol. 1, *In the School of Anti-Slavery, 1840 to 1866* (New Brunswick, NJ: Rutgers University Press, 1997), 402–3.
10. Quoted in Lewis Perry, *Boats against the Current: American Culture between Revolution and Modernity, 1820–1860* (New York: Oxford University Press, 1993), 60.
11. A Gentleman, *The Laws of Etiquette or, Short Rules and Reflections for Conduct in Society*, new ed. (Philadelphia: Carey, Lea & Blanchard, 1838), 208, AAS.
12. Peter Parley, *What to Do, and How to Do It; Or, Morals and Manners Taught by Examples* (New York: Wiley & Putnam, 1844), 99.

13. Herman Melville, *Moby-Dick; or, The Whale* (New York: Harper and Brothers, 1851), 186.

14. Quoted in Mimosa Stephenson, "Humor in *Uncle Tom's Cabin*," *Studies in American Humor*, new ser., 3 (2009), 13.

15. *Advocate of Moral Reform and Family Guardian* 18 (October 15, 1852), 157.

16. Letter from Dolley Payne Todd Madison to Francis (Fanny) Dandridge Henley Lear, April 27, 1833, *The Papers of Dolley Madison Digital Edition*, ed. Holly C. Shulman (Charlottesville: University of Virginia Press, Rotunda, 2008), http://rotunda.upress.virginia.edu/founders/DYMN-01-04-02-0190.

17. Caroline Healey Dall, *Selected Journals of Caroline Healey Dall*, vol. 1, *1838–1855*, ed. Helen R. Deese (Charlottesville: University of Virginia Press, 2006), 29.

18. George Templeton Strong, *Diary*, vol. 1, *Young Man in New York, 1835–1849*, ed. Allan Nevins and Milton Halsey Thomas (New York: Macmillan, 1952), 171.

19. Frederick William Seward, *Reminiscences of a War-Time Statesman and Diplomat, 1830–1915* (New York: G. P. Putnam's Sons, 1916), 58.

20. October 14, 1843, Dexter Russel Wright Papers, LoC.

21. E. Cleveland to A. Gilmore, December 13, 1845, folder labeled "Letters to Addison Gilmore, 1845, Mostly Concord RR," case 5, BARC.

22. Leslie Wheeler, ed., *Loving Warriors: Selected Letters of Lucy Stone and Henry B. Blackwell, 1853 to 1893* (New York: Dial, 1981), 34.

23. Philip N. Racine, *Gentlemen Merchants: A Charleston Family's Odyssey, 1828–1870* (Knoxville: University of Tennessee Press, 2008), 175.

24. Constance Rourke, *American Humor: A Study of the National Character* (1931; repr., New York: New York Review of Books, 2004), 237–38. For the early development of these almanacs, see Marion Barber Stowell, "Humor in Colonial Almanacs," *Studies in American Humor* 3 (1976), 34–47.

25. *Gems of American Wit and Anecdote; Or, the American Joe Miller* (London: Charles Tilt, 1839), 18–19, 76, 88.

26. H. Hastings Weld, "The Martyr to Science," *Boston Pearl* (October 1835), 46, 47.

27. *Rip Snorter Comic Almanac, 1859* (New York: Philip J. Cozans, [1858]).

28. *Finn's Comic Almanac, or United States Calendar, for 1835* (Boston: Marsh, Capen & Lyon, n.d.).

29. *NE* (December 10, 1857).

30. *The Old American Comic Almanac, 1843* (Boston: Thomas Groom & Co., n.d.), 28.

31. *NE* (November 20, 1856).

32. *The Ball of Yarn, or Queer, Quaint and Quizzical Stories, Unraveled. With Nearly 200 Comic Engravings* (New York: Elton, 1854), AAS.

33. *Fisher's Comic Almanac, 1845* (Boston: James Fisher, n.d.).

34. *Turner's Comic Almanack, 1839* (New York: Turner & Fisher, n.d.).

35. *Carpet-Bag* (October 16, 1852), front cover.

36. *Elton's Comic All-My-Nack, 1835* (New York: R. H. Elton, 1835).

37. *American Comic Almanac, 1836: With Whims, Scraps and Oddities* (Philadelphia: Grigg & Elliot, n.d.), 18. For crucial historical background of the trope of annihilating space, see Alexis D. Litvine, "The Annihilation of Space: A Bad (Historical) Concept," *Historical Journal* 65 (2022), 874–75.

38. *Gems of American Wit*, 110.

39. *Old American Comic Almanac, 1839* (Boston: S. N. Dickinson, n.d.). According to Vanessa Meikle Schulman, depictions of skeletons driving trains and other grisly imagery "peaked in the late 1850s and early 1860s, years coinciding with the highest rates of per capita railroad-related mortality and multiple-fatality accidents in the United States." Vanessa Meikle Schulman, *Work Sights: The Visual Culture of Industry in Nineteenth-Century America* (Amherst: University of Massachusetts Press, 2015), 38. For more on gallows humor, see Robert Gudmestad, *Steamboats and the Rise of the Cotton Kingdom* (Baton Rouge: Louisiana State University Press, 2011), 112.

40. *Rough and Ready Jester, Being a Funny Collection of Anecdotes, Witticisms, and Odd Sayings* (New York: C. P. Huestis, n.d.), 36, AAS.

41. *Young America's Comic Almanac for 1857* (New York: T. W. Strong, n.d.).

42. James H. Justus, *Fetching the Old Southwest: Humorous Writing from Longstreet to Twain* (Columbia: University of Missouri Press, 2004), 314–15.

43. *CA* (September 26, 1840).

44. *NE* (November 4, 1847).

45. William T. Porter, ed., *The Big Bear of Arkansas, and Other Sketches, Illustrative of Characters and Incidents in the South and South-West* (Philadelphia: Henry C. Baird, 1850), 106–12 (quotation on 112).

46. *NE* (September 1, 1859).

47. Cameron C. Nickels, *New England Humor: From the Revolutionary War to the Civil War* (Knoxville: University of Tennessee Press, 1993), 107–8, 116.

48. Samuel Putnam Avery, *The Book of 1000 Comical Stories* (New York: Dick & Fitzgerald, 1859), 105.

49. *Princeton Magazine* (January 1850), 229–33.

50. While it is difficult to know the exact number of privately owned pianos in the antebellum era, Arthur Loesser estimates that piano ownership increased during the antebellum era, with growing domestic production adding to foreign imports. Arthur Loesser, *Men, Women, and Pianos: A Social History*, new ed. (New York: Dover, 1990), 469, 492.

51. Charlotte N. Eyerman and James Parakilas, "1820s to 1870s: The Piano Calls the Tune," in *Piano Roles: Three Hundred Years of Life with the Piano*, ed. James Parakilas (New Haven, CT: Yale University Press, 1999), 190, 200.

52. Billy Coleman, *Harnessing Harmony: Music, Power, and Politics in the United States, 1788–1865* (Chapel Hill: University of North Carolina Press, 2020), 7.

53. "Carrollton March," composed by A. Clifton (Baltimore: John Cole, 1828). For an overview of this early music, see Norm Cohen, *Long Steel Rail: The Railroad in American Folksong* (Urbana: University of Illinois Press, 1981), chap. 3.

54. James D. Dilts, *The Great Road: The Building of the Baltimore & Ohio, the Nation's First Railroad, 1828–1853* (Stanford, CA: Stanford University Press, 1993), 10–11.

55. "Railroad March," composed by C. Meineke (Baltimore: George Willig, 1828).

56. "Railroad Quick Step," composed by W. Broadbent (Baltimore: George Willig, 1828).

57. "The Rail Road," composed by C. Meineke (Baltimore: John Cole, 1828).

58. As Dieter W. Hopkin notes of imagery of railroads in Europe during the same time period, "Representations . . . varied from the technically accurate to the weird and

wonderful." Dieter W. Hopkin, "Reflections on the Iconography of Early Railways," in *Early Railways: A Selection of Papers from the First International Early Railways Conference*, ed. Andy Guy and Jim Rees (London: Newcomen Society, 2001), 347. See also Albert Churella, *The Pennsylvania Railroad*, vol. 1, *Building an Empire, 1846–1917* (Philadelphia: University of Pennsylvania Press, 2013), 45, for another image that indicates unfamiliarity with trains.

59. Such as "Rangers' Trip to Westborough," composed by James Hooton (Boston: C. Bradlee, 1834), and the "Rail Road Waltz," composed by Ch. Zeuner (Boston: Oliver Ditson, 1835). For additional titles, see John Ashton & Co., "Catalogue of vocal and instrumental music," between 1834 and 1844, AAS-B.

60. "Alsacian Rail Road Gallops," composed to J. Guignard (Philadelphia: A. Fiot, 1845).

61. "Empire Quick Step," composed by J. Long (Boston: Oliver Ditson, 1844).

62. "Knickerbocker Schottische," composed by Charles Steinruck (New York: Horace Waters, 1851).

63. "Harnden's Express Line Gallopade and Trio," composed by A. R. (New York: Firth and Hall, 1841).

64. "New Orleans and Great Northern Railroad Polka," composed by Theodore La Hache (New York: Firth, Pond, and Co., 1854).

65. "Erie Railroad Polka," composed by Van Der Weyde (New York: William Vanderbeek, 1851).

66. "Fast Line Gallop," composed by James N. Beck (Philadelphia: Lee and Walker, 1853).

67. "Chatwa Pic Nic Mazurka," composed by T. S. H. (New Orleans: Ph. P. Werlein, 1856).

68. "Clear the Way!," composed by Stephen C. Massett (San Francisco: Stephen Massett, 1856).

69. Silas Redington, "Rail road song," 1847, AAS-B.

70. "Ridin' in a Rail Road Keer," composed by W. J. Florence (St. Louis: R. J. Compton, 1859).

71. *Free Soil Minstrel* (New York: Martyn & Elyn, 1848), 205, AAS.

72. "Underground Rail Car," composed by George N. Allen (Cleveland: S. Brainerd, 1854).

73. "The Ghost of Uncle Tom," composed by Martha Hill (New York: Horace Waters, 1854).

74. "Get Off the Track!," composed by Jesse Hutchinson, Jr. (Boston: Jesse Hutchinson, Jr., 1844).

75. Charles E. Hamm, *Yesterdays: Popular Song in America* (New York: W. W. Norton, 1979), 150; Brian Roberts, *Blackface Nation: Race, Reform, and Identity in American Popular Music, 1812–1925* (Chicago: University of Chicago Press, 2017), chap. 8.

76. For American concert practice during this time, see Daniel Cavicchi, *Listening and Longing: Music Lovers in the Age of Barnum* (Middletown, CT: Wesleyan University Press, 2011), chap. 1.

77. Dall, *Selected Journals*, 1:255.

78. "Last night of the Harmoneons grand vocal and instrumental soiree," 1848, AAS-B.

79. "Flagg's Hall Worcester," 1851, AAS-B.

80. Roger L. Beck and Richard K. Hansen, "Josef Gungl and His Celebrated American Tour: November 1848 to May 1849," *Studia Musicologica Academiae Scientiarum Hungaricae* 36 (1995), 55–56, 67–68.

81. Joseph J. Mersman, *The Whiskey Merchant's Diary: An Urban Life in the Emerging Midwest*, ed. Linda A. Fisher (Athens: Ohio University Press, 2007), 68–69.

82. "Grand instrumental concert at the Melodeon," 1848, AAS-B; *New York Herald* (December 26, 1847); Washington, DC, *Daily Union* (February 18, 1848); New Orleans *Daily Crescent* (April 10, 1848).

83. *New York Herald* (January 23, June 19, and June 22, 1848).

84. Washington, DC, *Daily Union* (March 2, 1849).

85. "Grand instrumental concert by the celebrated Saxonia Orchestra," 1849, AAS-B. For another performance of the piece by different groups, see "Farewell concert, and positively the last!," 1851, AAS-B; and *New York Herald* (November 29, 1854).

86. Lawrence W. Levine, *Highbrow/Lowbrow* (Cambridge, MA: Harvard University Press, 1990), 111–12; Nancy Newman, *Good Music for a Free People: The Germania Musical Society in Nineteenth-Century America* (Rochester, NY: University of Rochester Press, 2010), 122.

87. "City Hall: May day evening," 1850, AAS-B. The group performed identical material on May 8, 1850. "Westminster Hall," 1850, AAS-B. For a list of the Germania Musical Society's performances, see Newman, *Good Music*, appendix A. For advertisements of additional performances, see Alexandria (Virginia) *Gazette* (January 22, 1851) and Washington, DC, *Republic* (March 11, 1851).

88. Cavicchi, *Listening and Longing*, 32.

89. "Melodeon! Monday evening, Feb. 12th, 1855," 1855, AAS-B. For another performance by the same group, also including the railroad piece, see "Last night at Brinley Hall!," 1855, AAS-B.

90. Nashville *Daily Patriot* (March 19, 1856).

91. "City Hall! King's Æolians! Great attraction!," 1858, AAS-B.

92. Schulman, *Work Sights*, 2–9. This chapter will focus on imagery in popular culture; for images of the railroad in art and photography, see Ian Kennedy, "Crossing Continents: America and Beyond," in *The Railway: Art in the Age of Steam*, ed. Ian Kennedy and Julian Treuherz (New Haven, CT: Yale University Press, 2008), 119–53.

93. For a general discussion of Salt River, see Liz Hutter, "'Ho for Salt River!': Politics, Loss, and Satire," *Common-place* 7 (April 2007), http://commonplace.online/article/ho-for-salt-river/.

94. Bryan F. Le Beau, *Currier and Ives: America Imagined* (Washington, DC: Smithsonian Institution Press, 2001), 298.

95. Stanley L. Baker and Virginia Brainard Kunz, *The Collector's Book of Railroadiana* (New York: Hawthorne Books, 1976), 16.

96. Karin Hertel McGinnis, "Moving Right Along: Nineteenth Century Panorama Painting in the United States" (Ph.D. diss., University of Minnesota, 1983), chap. 5.

97. John Francis McDermott, *The Lost Panoramas of the Mississippi* (Chicago: University of Chicago Press, 1958), 44.

98. McDermott, *Lost Panoramas of the Mississippi*, 53, 61–62.

99. *Phelps's Travellers' Guide through the United States; Containing upwards of Seven Hundred Rail-road, Canal, and Stage and Steam-boat Routes, Accompanied with a New Map of the United States* (New York: Ensigns & Thayer, 1848).

100. *Mitchell's New Traveller's Guide through the United States, Containing the Principal Cities, Towns, &c. Alphabetically Arranged; together with the Railroad, Stage, Steamboat and Canal Routes, with the Distances, in Miles, from Place to Place* (Philadelphia: Thomas, Cowperthwait & Co., 1849).

101. Will Mackintosh, *Selling the Sights: The Invention of the Tourist in American Culture* (New York: New York University Press, 2019), 24, 36, 44, 48; Richard H. Gassan, *The Birth of American Tourism: New York, the Hudson Valley, and American Culture, 1790–1830* (Amherst: University of Massachusetts Press, 2008), 71, 77. Carolyn Kitch suggests that in the 1860s, far more people purchased guides for cross-continental trips than actually engaged in cross-continental travel. Carolyn Kitch, "'A Piazza from Which the View Is Constantly Changing': The Promise of Class and Gender Mobility on the Pennsylvania Railroad's Cross-Country Tours," *Pennsylvania History* 72 (2005), 508. For the development of railroad timetables and the reading skills necessary to decipher them, see Mike Esbester, "Designing Time: The Design and Use of Nineteenth-Century Transport Timetables," *Journal of Design History* 22, no. 2 (2009), 91–113; and Mike Esbester, "Nineteenth-Century Timetables and the History of Reading," *Book History* 12 (2009), 156–85.

102. *Horn's Railroad Gazette* (June 9, 1849).

103. John Disturnell, *The Traveller's Guide through the State of New York, Canada, &c: Embracing a General Discription of the City of New-York; The Hudson River Guide, and the Fashionable Tour to the Springs and Niagara Falls; with Steam-boat, Rail-road, and Stage Routes; Accompanied by Correct Maps* (New York: J. Disturnell, 1836), 25, 32.

104. *Guide to the Lakes and Mountains of New-Hampshire, via the Several Routes Connecting with the Boston, Concord & Montreal Railroad at Concord, N.H.* (Concord, NH: Tripp & Osgood, 1852), 18, AAS.

105. *The Western Traveler's Pocket Directory and Stranger's Guide* (Schenectady, NY: S. S. Riggs, 1834); and *The Ohio Railroad Guide: Illustrated and Descriptive. Part I. Cincinnati, Hamilton & Dayton Railway, and Mad River and Lake Erie Railway from Dayton to Springfield* (Cincinnati: Cincinnati Gazette Company, 1852).

106. *Hudson River and the Hudson River Railroad: With a Complete Map, and Wood Cut Views of the Principal Objects of Interest upon the Line* (Boston: Bradbury & Guild, 1851); and Pennsylvania Railroad, *Guide for the Pennsylvania Railroad, with an Extensive Map: Including the Entire Route, with All Its Windings, Objects of Interest, and Information Useful to the Traveler* (Philadelphia: T. K. & P. G. Collins, 1855).

107. Eli Bowen, *Off-Hand Sketches; A Companion for the Tourist and Traveller over the Philadelphia, Pottsville, and Reading Railroad. Describing the Scenery, Improvements, Mineral and Agricultural Resources, Historical Incidents, and Other Subjects of Interest in the Vicinity of the Route. With Numerous Engravings* (Philadelphia: J. W. Moore, 1854), 10.

108. Charles P. Dare, *Philadelphia, Wilmington and Baltimore Railroad Guide: Containing a Description of the Scenery, Rivers, Towns, Villages, and Objects of Interest along the Line of Road: Including Historical Sketches, Legends, &c.* (Philadelphia: Fitzgibbon & Van Ness, 1856), 23.

109. Brad S. Lomazzi, *Railroad Timetables, Travel Brochures & Posters: A History and Guide for Collectors* (Spencertown, NY: Golden Hill, 1995), 74.

110. W. Williams, *Appleton's New and Complete United States Guide Book for Travellers: Embracing the Northern, Eastern, Southern, and Western States, Canada, Nova Scotia, New Brunswick, etc.* (New York: D. Appleton, 1850), 3, 5.

111. *Harper's New York and Erie Rail-Road Guide Book: Containing a Description of the Scenery, Rivers, Towns, Villages, and Most Important Works on the Road* (New York: Harper & Brothers, 1851), 103.

112. *Illustrated American News* (November 1851), 2, 6.

113. William Guild, *New York and the White Mountains; With a Complete Map, and Numerous Wood-Cut Views of the Principal Objects of Interest upon the Line* (Boston: Bradbury & Guild, 1852), 5, AAS.

114. Edward Deering Mansfield, *The Ohio Railroad Guide, Illustrated: Cincinnati to Erie via Columbus and Cleveland* (Columbus: Ohio State Journal Company, 1854), 3, 4–5.

115. Joseph H. Bragdon, *Seaboard Towns; Or, Traveller's Guide Book from Boston to Portland: Containing a Description of the Cities, Towns and Villages, Scenery and Objects of Interest along the Route of the Eastern Railroad and Its Branches, and the Portland, Saco and Portsmouth Railroad. Including Historical Sketches, Legends, &c.* (Newburyport, MA: Moulton & Clark, 1857), 53.

Chapter 4 • Religion

1. Daniel W. Howe, *What Hath God Wrought: The Transformation of America, 1815–1848* (New York: Oxford University Press, 2007), 186.

2. John Hedley Brooke, *Science and Religion: Some Historical Perspectives* (1991; repr., Cambridge: Cambridge University Press, 2014), 56.

3. Peter Harrison, "'Science' and 'Religion': Constructing the Boundaries," in *Science and Religion: New Historical Perspectives*, ed. Thomas Dixon, Geoffrey Cantor, and Stephen Pumfrey (New York: Cambridge University Press, 2010), 26 (quotation); Frank M. Turner, "The Late Victorian Conflict of Science and Religion as an Event in Nineteenth-Century Intellectual and Cultural History," in *Science and Religion: New Historical Perspectives*, ed. Thomas Dixon, Geoffrey Cantor, and Stephen Pumfrey (New York: Cambridge University Press, 2010), 87–89; Peter Harrison, "Religion, Scientific Naturalism and Historical Progress," in *Religion and Innovation: Antagonists or Partners?*, ed. Donald A. Yerxa (London: Bloomsbury, 2016), 87–88; Frederick Ferre, *Philosophy of Technology* (Athens: University of Georgia Press, 1995), 99; and David E. Nye, *America as Second Creation: Technology and Narratives of New Beginnings* (Cambridge, MA: MIT Press, 2003), 10, 40.

4. Thomas P. Hughes, *Human-Built World: How to Think about Technology and Culture* (Chicago: University of Chicago Press, 2004), 30.

5. David Paul Nord, *Faith in Reading: Religious Publishing and the Birth of Mass Media in America* (New York: Oxford University Press, 2004), 114.

6. Presbyterians split in 1837; Methodists in 1843; and Baptists in 1845. James H. Moorhead, "The 'Restless Spirit of Radicalism': Old School Fears and the Schism of 1837," *Journal of Presbyterian History* 78 (2000), 19; Chris Padgett, "Hearing the Antislavery Rank-and-File: The Wesleyan Methodist Schism of 1843," *Journal of the Early Republic* 12 (1992), 82; and Glen Jeansonne, "Southern Baptist Attitudes toward Slavery, 1845–1861," *Georgia Historical Quarterly* 55 (1971), 510.

7. Horace Bushnell, *The Day of Roads: A Discourse, Delivered on the Annual Thanksgiving, 1846* (Hartford, CT: Elihu Geer, 1846), 3, 7, 9, 31, 35. The prophecy is from Isaiah 45:2, KJV.

8. S. C. Aiken, *Moral View of Rail Roads. A Discourse, Delivered on Sabbath Morning, February 23, 1851, on the Occasion of the Opening of the Cleveland and Columbus Rail Road* (Cleveland: Harris, Fairbanks & Co., 1851), 12, 17, AAS.

9. For example, see *Augusta* (Georgia) *Chronicle* (March 22, 1851), *Boston Recorder* (March 27, 1851), Schenectady (NY) *Cabinet* (March 18, 1851), and the Milwaukee *Wisconsin Free Democrat* (April 9, 1851). The full text from Nahum is "The chariots shall rage in the streets, they shall justle one against another in the broad ways: they shall seem like torches, they shall run like the lightnings." Nahum 2:4, KJV.

10. Joseph G. Wilson, *The Pacific Rail Road: A Discourse, Delivered in the Second Presbyterian Church, of Lafayette, on Thanksgiving Day, November 28, 1850* (Lafayette, IN: Rosser & Brother, 1850), 4, 17, 21, AAS.

11. *Christian Visitant* (November 1827), 294.

12. *A Treatise on the Millennium, Shewing Its Near Approximation, Especially by the Accomplishment of Those Events Which Were to Precede It; The Second Advent or Coming of Our Lord and Saviour Jesus Christ; and the Restoration of a State of Paradise upon Earth* (Boston: Printed for the Author, 1838), 236, AAS.

13. Alexander W. Bradford, *A Discourse Delivered before the New York Historical Society, at Its Forty-First Anniversary, 20th November, 1845* (New York: Press of the Historical Society, 1846), 26–27, AAS.

14. *Monthly Religious Magazine* (October 1851), 471, 480.

15. *Princeton Review* (January 1850), 132.

16. Joseph A. Copp, *The Atlantic Telegraph: As Illustrating the Providence and Benevolent Designs of God. A Discourse, Preached in the Broadway Church, Chelsea, August 8, 1858* (Boston: T. R. Marvin & Son, 1858), 14, AAS.

17. "Conversion of the Heathen a Difficult Work," *American Quarterly Register* (August 1836), 69.

18. *CA* (December 8, 1838).

19. Charlotte Hixon, "Trip to Aiken," May 28, 1849, SCL.

20. *The New York Comic Almanac for 1857* (New York: Philip J. Cozans, n.d.).

21. John F. Pollard, *Money and the Rise of the Modern Papacy: Financing the Vatican, 1850–1950* (Cambridge: Cambridge University Press, 2005), 29; Owen Chadwick, *A History of the Popes, 1830–1914* (Oxford: Clarendon Press, 1998), 50–51, 61. For a general overview of anti-Catholicism in the antebellum United States, see Jon Gjerde, *Catholicism and the Shaping of Nineteenth-Century America*, ed. S. Deborah Kang (New York: Cambridge University Press, 2012).

22. N. Rounds, "A Lecture on Education," *Methodist Magazine* (July 1837), 268.

23. H. P. Tappan, "The Immaculate Conception of the Virgin," *American Protestant Magazine* (March 1848), 299.
24. Gjerde, *Catholicism*, 37.
25. *Sunday School Journal* (June 1849), 101 (quotation); Gjerde, *Catholicism*, 44.
26. "The Railroad Jubilee," *Well-Spring* (October 1851), 167.
27. "Walking Spanish," *Well-Spring* (September 1852), 154.
28. Anna Ella Caroll, *The Romish Church Opposed to the Liberties of the American People. With a Biographical Sketch of the Hon. Erastus Brooks; His Celebrated Controversy with Arch-Bishop Hughes, &c.* (Boston: James French and Company, 1856), 8, AAS.
29. Gjerde, *Catholicism*, 88.
30. "Notes By the Way. No. VIII," *Sunday School Visiter* (August 1836), 262.
31. *CA* (November 25, 1837).
32. *CA* (March 11, 1837).
33. "A Sermon on a Steamboat," *Christian Watchman* (December 5, 1845).
34. *American Messenger* (January 1843), 7.
35. *First Annual Report of the Church Extension Committee of the General Assembly of the Presbyterian Church in the United States of America, Presented May, 1856, and the Church Extension Summary, or a Brief View of the Nature and Importance of the Church Extension Enterprise* (St. Louis: Sherman Spencer, 1856), 15, AAS.
36. *NE* (September 17, 1857).
37. Baltimore *Sun* (April 15, 1857).
38. Lowell (Massachusetts) *Daily Citizen* (June 13, 1859).
39. Frederick Plummer Diary, AAS.
40. "What One Man May Do," *American Messenger* (May 1851), 19.
41. John Lardas Modern, *Secularism in Antebellum America, with Reference to Ghosts, Protestant Subcultures, Machines, and Their Metaphors; Featuring Discussions of Mass Media, Moby-Dick, Spirituality, Phrenology, Anthropology, Sing Sing State Penitentiary, and Sex with the New Motive Power* (Chicago: University of Chicago Press, 2011), 95, 28 (quotation).
42. John L. Brooke, "Cultures of Nationalism, Movements of Reform, and the Composite-Federal Polity: From Revolutionary Settlement to Antebellum Crisis," *Journal of the Early Republic* 29 (2009), 21.
43. H. M. Saxon, "Distribution of New Church Books," *Age* (November 1852), 137–38.
44. George Winfred Hervey, *The Principles of Courtesy: With Hints and Observations on Manners and Habits* (New York: Harper & Brothers, 1852), 198–200.
45. Alexander Mackay, *The Western World; Or, Travels in the United States in 1846–47: Exhibiting Them in Their Latest Development, Social, Political, and Industrial; Including a Chapter on California. With a New Map of the United States, Showing Their Recent Territorial Acquisitions, and a Map of California*, vol. 1 (1849; repr., New York: Negro Universities Press, 1968), 31.
46. N. Murray, "Causes of the Present Declension of Religion," *Literary and Theological Review* 12 (December 1836), 491–92.
47. "Editor's Note. Victims of Progress," *Harper's* (November 1852), 839–40.
48. Broadside dated December 19, 1838, folder labeled "Western Railroad Administrative Papers, 1829–1839," case 3, BARC.

49. *Boston Recorder* (April 26, 1839).

50. L. F. Dimmick, *A Discourse on the Moral Influence of Rail-Roads* (Boston: Tappan & Dennet, 1841), 12–13, 101.

51. "Influence of Railroads," San Francisco *Pacific* (January 1, 1852).

52. "Monthly Concert," *Religious Intelligencer* (September 5, 1835), 210–11.

53. *Bridgeton* (New Jersey) *Chronicle* (October 26, 1839).

54. O. A. Brownson, *Social Reform. An Address before the Society of the Mystical Seven in the Wesleyan University, Middleton, Conn. August 7, 1844* (Boston: Waite, Peirce & Company, 1844), 34, AAS.

55. J. G. Adams, "Home, and the Sabbath School," *Gospel Teacher* 9 (October / November 1847), 61.

56. Robert H. Abzug, *Cosmos Crumbling: American Reform and the Religious Imagination* (New York: Oxford University Press, 1994), 112; Richard R. John, "Taking Sabbatarianism Seriously: The Postal System, the Sabbath, and the Transformation of American Political Culture," *Journal of the Early Republic* 10 (1990), 517–67; and Alexis McCrossen, *Holy Day, Holiday: The American Sunday* (Ithaca, NY: Cornell University Press, 2000).

57. "The General Camp-Meeting," *Church Advocate* (September 4, 1852), 146.

58. "Remember the Sabbath Day to Keep It Holy," *Bethel* (November 1835), 133.

59. *Essex North* (Newburyport, Massachusetts) *Register* (September 23, 1836).

60. Letter from J. Obear to Mrs. Obear, February 17, 1849, Obear Family Papers, SCL.

61. Entry for March 29, 1852, Esther Belle Hanna Diary, HM 31176, HL.

62. "Sabbath Movement," (Boston) *Sheet Anchor* (November 1844), 162.

63. "Hunt—Commerce, &c.," (Mobile) *Alabama Planter* (November 15, 1852).

64. *LR* (March 1841), 95.

65. "Sabbath Convention," *Gospel Publisher and Journal of Useful Knowledge* (April 1844), 109.

66. *NE* (June 21, 1855).

67. McCrossen, *Holy Day*, 26.

68. *Prohibition of Sunday Travelling on the Pennsylvania Rail Road* (Philadelphia: Merrihew and Thompson, 1850), 5.

69. "Beauties of the Sunday Law," political cartoon, LCP.

70. "Sunday Laws," political cartoon, LCP.

71. *Youth's Penny Gazette* 8 (January 30, 1850), 9.

72. Amy Mitchell-Cook, *A Sea of Misadventures: Shipwreck and Survival in Early America* (Columbia: University of South Carolina Press, 2013), 52.

73. Jay Newman, *Religion and Technology: A Study in the Philosophy of Culture* (Westport, CT: Praeger, 1997), 119.

74. Thomas Smyth, *The Voice of God in Calamity: Or, Reflections on the Loss of the Steam-Boat Home, October 9, 1837. A Sermon: Delivered in the Second Presbyterian Church, Charleston, on Sabbath Morning, October 22, 1837*, 2d ed. (Charleston, SC: Jenkins and Hussey, 1837), 9, 13, 15–16, 20.

75. Drew Gilpin Faust, *This Republic of Suffering: Death and American Civil War* (New York: Knopf, 2008), 19.

76. *A Full and Particular Account of All the Circumstances Attending the Loss of the Steamboat Lexington, in Long-Island Sound, on the Night of January 13, 1840; As Elicited in the Evidences of the Witnesses Examined Before the Jury of Inquest, Held in New-York Immediately After the Lamentable Event* (Providence, RI: H. H. Brown and A. H. Stilwell, 1840), unnumbered page and back cover, HL.

77. John S. Stone, *A Sermon, Occasioned by the Burning of the Steamer Lexington. Preached in St. Paul's Church, Boston* (Boston: Perkins & Marvin, 1840), 13–14, 16.

78. *A Memorial of Sa-Sa-Na, the Mohawk Maiden; Who Perished in the Rail Road Disaster at Deposit, N.Y., February 18, 1852* (Hamilton, NH: Waldron & Baker, 1852), 18.

79. Anna Marie Resseguie, *A View from the Inn: The Journal of Anna Marie Resseguie, 1851–1867* (Ridgefield, CT: Keeler Tavern Preservation Society, 1993), 78, 109–10.

80. D. R. Brewer, *The Loss of the San Francisco: Its Religious Lessons. A Sermon, Preached in Trinity Church, Newport, R. I., on Sunday, January 22d, 1854* (Providence, RI: George H. Whitney, 1854), 13, 14, AAS.

81. Entry for January 17, 1854, journal 5, Caroline Barrett White Journals, AAS.

82. F. Reck Harbaugh, *The Burlington Disaster. A Sermon Preached in the Presbyterian Church, in Burlington, N.J., Sunday, September 9, 1855, by F. Reck Harbaugh, Pastor* (Philadelphia: Henry B. Ashmead, 1855), 16, 9, 20, HL.

83. Henry A. Boardman, *God's Providence in Accidents. A Sermon Occasioned by the Deaths of the Rev. John Martin Connell, Mr. John Field Gillespie, and Mrs. Susan Gillespie, Three of the Victims of the Railroad Catastrophe at Burlington, New Jersey, on the 29th day of August, 1855* (Philadelphia: Parry and McMillan, 1855), 30.

84. Lemuel Porter, *Friendly Advice to Railroad Men. A Discourse Occasioned by the Death of Mr. Charles W. Nichols, an Engineer on the Western Railroad. Delivered in the Baptist Church, Pittsfield, Mass., September 7, 1856, by Rev. Lemuel Porter, D. D.* (Springfield, MA: Samuel Bowles & Company, 1856), 3, AAS.

85. J. B. Shaw, *Sermon Preached at the Brick Church, Rochester, on the Occasion of the Funeral of John Snell, Rail Road Engineer, on Sunday Afternoon, February 22, 1857, with an Editorial Notice of the Funeral from the Rochester Union and Advertiser* (Rochester, NY: Curtis, Butts & Co., 1857), 5, AAS.

86. Faust, *This Republic of Suffering*, 30–31.

Chapter 5 • Black Passengers

1. Frederick Douglass, *The Frederick Douglass Papers*, series 2, *Autobiographical Writings*, vol. 3, *Life and Times of Frederick Douglass*, book 1, *The Text and Editorial Apparatus*, ed. John R. McKivigan (New Haven, CT: Yale University Press, 2012), 76.

2. Susan Eva O'Donovan, "Thinking about the Political Lives of Slaves," *American Nineteenth Century History* 21 (2020), 28, 31.

3. Wiebe E. Bijker, *Of Bicycles, Bakelites, and Bulbs: Toward a Theory of Sociotechnical Change* (Cambridge, MA: MIT Press, 1995).

4. Aaron W. Marrs, *Railroads in the Old South: Pursuing Progress in a Slave Society* (Baltimore: Johns Hopkins University Press, 2009), chap. 3; Aaron Hall, "Slaves of the State: Infrastructure and Governance through Slavery in the Antebellum South," *Journal of American History* 106 (2019), 19–46; and Calvin Schermerhorn, *Money over*

Mastery, Family over Freedom: Slavery in the Antebellum Upper South (Baltimore: Johns Hopkins University Press, 2011), chap. 5. For enslaved people creating infrastructure and then using it for their own purposes, see Ryan A. Quintana, "Planners, Planters, and Slaves: Producing the State in Early National South Carolina," *Journal of Southern History* 81 (2015), 99.

5. "Recent Defences of Slavery," *Massachusetts Quarterly Review* (1849), 497.

6. James Gillespie Birney, *Letters of James Gillespie Birney, 1831–1857*, vol. 1, ed. Dwight L. Dumond (New York: Appleton, 1938), 411.

7. See, for example, Edlie L. Wong, *Neither Fugitive nor Free: Atlantic Slavery, Freedom Suits, and the Legal Culture of Travel* (New York: New York University Press, 2009); Blair L. M. Kelley, *Right to Ride: Streetcar Boycotts and African American Citizenship in the Era of* Plessy v. Ferguson (Chapel Hill: University of North Carolina Press, 2010); Graham Russell Gao Hodges, *David Ruggles: A Radical Black Abolitionist and the Underground Railroad in New York City* (Chapel Hill: University of North Carolina Press, 2010); R. J. M. Blackett, *Making Freedom: The Underground Railroad and the Politics of Slavery* (Chapel Hill: University of North Carolina Press, 2013); Kelly Kennington, "Law, Geography, and Mobility: Suing for Freedom in Antebellum St. Louis," *Journal of Southern History* 80 (2014), 575–604; Eric Foner, *Gateway to Freedom: The Hidden History of the Underground Railroad* (New York: W. W. Norton, 2015); Elizabeth Stordeur Pryor, *Colored Travelers: Mobility and the Fight for Citizenship before the Civil War* (Chapel Hill: University of North Carolina Press, 2016); R. J. M. Blackett, *The Captive's Quest for Freedom: Fugitive Slaves, the 1850 Fugitive Slave Law, and the Politics of Slavery* (New York: Cambridge University Press, 2018); Paul Finkelman, *Supreme Injustice: Slavery in the Nation's Highest Court* (Cambridge, MA: Harvard University Press, 2018); S. Charles Bolton, *Fugitivism: Escaping Slavery in the Lower Mississippi Valley, 1820–1860* (Fayetteville: University of Arkansas Press, 2019); Robert H. Churchill, *The Underground Railroad and the Geography of Violence in Antebellum America* (New York: Cambridge University Press, 2020); O'Donovan, "Thinking about the Political Lives," 25–37; and Timothy D. Walker, ed., *Sailing to Freedom: Maritime Dimensions of the Underground Railroad* (Amherst: University of Massachusetts Press, 2021).

8. Marrs, *Railroads in the Old South*.

9. See, for example, the following issues of *CA*: April 22 and 29, 1837; January 13, March 3, April 19, July 28, and August 18, 1838; February 16, 1839; and May 1, 1841.

10. Schermerhorn, *Money over Mastery*, 174–80.

11. Anthony Kaye, "The Second Slavery: Modernity in the Nineteenth-Century South and the Atlantic World," *Journal of Southern History* 75 (2009), 633.

12. A. J. McElveen, *Broke by the War: Letters of a Slave Trader*, ed. Edmund L. Drago (Columbia: University of South Carolina Press, 1991), 46, 63.

13. Robert Colby, "'Observant of the Laws of this Commonwealth': A Free Black Family between Forced Migration and Slave Capitalism," *Journal of the Early Republic* 42 (2022), 347.

14. Entry for September 13, 1856, Edward Spann Hammond, Diary of a Trip to Cincinnati, SCL.

15. Bolton, *Fugitivism*, 106.

16. *Charleston Courier* (February 26, 1850).

17. Wilma A. Dunaway, "Put in Master's Pocket: Cotton Expansion and Interstate Slave Trading in the Mountain South," in *Appalachians and Race: The Mountain South from Slavery to Segregation*, ed. John C. Inscoe (Lexington: University Press of Kentucky, 2001), 117.

18. Damian Alan Pargas, *Slavery and Forced Migration in the Antebellum South* (New York: Cambridge University Press, 2014), 85–86, 95, 97.

19. Abraham Lincoln, *The Collected Works of Abraham Lincoln*, vol. 1, *1824–1848*, ed. Roy P. Basler (Brunswick, NJ: Rutgers University Press, 1953), 260.

20. Entry for April 4–5, 1857, Henry Ashworth diary, SCL.

21. *NS* (June 30, 1848).

22. *NS* (November 17, 1848).

23. *NE* (February 19, 1852).

24. *NE* (January 27, 1859).

25. Robert Gudmestad, *Steamboats and the Rise of the Cotton Kingdom* (Baton Rouge: Louisiana State University Press, 2011), 56.

26. Marrs, *Railroads in the Old South*, 198.

27. Damian Alan Pargas notes that forcing the enslaved people to walk had some advantages for slave traders, but trains and steamboat also had advantages, particularly for direct transportation to Natchez or New Orleans. Pargas, *Slavery and Forced Migration*, 45, 47.

28. *Douglass' Monthly* (April 1860).

29. *NE* (May 21, 1857).

30. Anonymous Travel Diary Transcript, p. 68, WHS (call number SC 894).

31. Douglas R. Egerton, "Markets without a Market Revolution: Southern Planters and Capitalism," *Journal of the Early Republic* 16 (1996), 264–65.

32. Timothy D. Walker, "Sailing to Freedom: Maritime Dimensions of the Underground Railroad," in *Sailing to Freedom: Maritime Dimensions of the Underground Railroad*, ed. Timothy D. Walker (Amherst: University of Massachusetts Press, 2021), 19.

33. Cassandra Newby-Alexander, "Hampton Roads and Norfolk, Virginia as a Waypoint and Gateway for Enslaved Persons Seeking Freedom," in *Sailing to Freedom: Maritime Dimensions of the Underground Railroad*, ed. Timothy D. Walker (Amherst: University of Massachusetts Press, 2021), 80–81.

34. Michael D. Thompson, *Working on the Dock of the Bay: Labor and Enterprise in an Antebellum Southern Port* (Columbia: University of South Carolina Press, 2015), 64, 76 (quotation); Kennington, "Law, Geography, and Mobility," 595; Matthew Salaria, *Slavery's Borderland: Freedom and Bondage along the Ohio River* (Philadelphia: University of Pennsylvania Press, 2013), chap. 3.

35. John Mangipano, "Social Geography of Interstate Escape: Runaway Slaves from Louisiana and Mississippi, 1800–1860," *Journal of Mississippi History* 75 (2013), 147; Bolton, *Fugitivism*, 172–73.

36. Quoted in the New York *Emancipator* (November 15, 1838).

37. New Orleans *Daily Picayune* (August 22, 1841).

38. Columbus *Ohio Statesman* (September 15, 1841).

39. Quoted in *CA* (September 4, 1841). For additional references to concern about free Black people, see the Baltimore *Sun* (September 1, 1841).

40. Bolton, *Fugitivism*, 127.

41. Walter Johnson, *River of Dark Dreams: Slavery and Empire in the Cotton Kingdom* (Cambridge, MA: Belknap Press of Harvard University Press, 2013), 144.

42. *Sandusky* (Ohio) *Register* (June 5, 1860).

43. *NE* (February 2, 1860).

44. "Sir, enclosed you will receive a copy of the proceedings . . . ," 1835, AAS-B.

45. *Alexandria* (Virginia) *Gazette* (September 7, 1836).

46. Bolton, *Fugitivism*, 177.

47. Johnson, *River of Dark Dreams*, 147.

48. David Schley, *Steam City: Railroads, Urban Space, and Corporate Capitalism in Nineteenth-Century Baltimore* (Chicago: University of Chicago Press, 2020), 53.

49. Philadelphia *Public Ledger* (July 7, 1840). For another account of the Louisiana law, see the Philadelphia *North American* (July 6, 1840). For an example of how this law dissuaded a sympathetic ship captain, see Bolton, *Fugitivism*, 72.

50. Bolton, *Fugitivism*, 115.

51. Paul Finkelman, *State Slavery Statutes: A Guide to the Microform Edition* (Frederick, MD: UPA Academic Editions, 1989), 261. The statute itself is available on the microform referenced on that page. I am grateful to Kathy Hilliard for directing me to this source.

52. *Alexandria* (Virginia) *Gazette* (August 17, 1841).

53. Megan Jeffreys, "Freedom on the Move by Sea: Evidence of Maritime Escape Strategies in American Runaway Slave Advertisements," in *Sailing to Freedom: Maritime Dimensions of the Underground Railroad*, ed. Timothy D. Walker (Amherst: University of Massachusetts Press, 2021), 207.

54. Thompson, *Working*, 80–84.

55. David S. Cecelski, "Black Watermen, Fugitives from Slavery, and an Old Woman on the Edge of a Swamp," in *Sailing to Freedom: Maritime Dimensions of the Underground Railroad*, ed. Timothy D. Walker (Amherst: University of Massachusetts Press, 2021), 71.

56. Finkelman, *State Slavery Statutes*, 87. The statute itself is available on the microform referenced on that page.

57. Baltimore *Sun* (March 23, 1850).

58. *New-York Daily Tribune* (July 14, 1856).

59. *Lynchburg* (Virginia) *Daily Virginian* (August 30, 1852).

60. *Report of the President and Directors of the East Tennessee and Georgia Rail Road Company, to the Stockholders in the Same, at their Annual Convention, Held at Athens, Sept. 3, 1856* (Athens, TN: Printed by Sam. P. Ivins, Office of the "Post," 1857), 30.

61. Quoted in *Columbus* (Georgia) *Tri-Weekly Enquirer* (January 10, 1856).

62. Quoted in *FDP* (September 15, 1854).

63. For one accounting of fugitive slave rescues, which suggests the scale of attempts, see Churchill, *Underground Railroad*, 234–42.

64. Henry Box Brown, *The Narrative of the Life of Henry Box Brown* (New York: Oxford University Press, 2002).

65. New York *Commercial Advertiser* (June 25, 1845).

66. *Alexandria* (Virginia) *Gazette* (October 21, 1848).

67. Blackett, *Captive's Quest for Freedom*, 145.

68. *NS* (May 25, 1849).

69. Cheryl Janifer Laroche, "The Underground Railroad in Maryland's Ports, Bays, and Harbors: Maritime Strategies for Freedom," in *Sailing to Freedom: Maritime Dimensions of the Underground Railroad*, ed. Timothy D. Walker (Amherst: University of Massachusetts Press, 2021), 109.

70. Blackett, *Making Freedom*, 68–69.

71. *Alexandria* (Virginia) *Gazette* (September 7, 1836).

72. *Douglass' Monthly* (April 1859). For another account, see *Bennington Banner* (April 1, 1859).

73. *NS* (July 20, 1849). For the Crafts's account of their story, see William Craft, *Running a Thousand Miles for Freedom* (London: William Tweedie, 1860).

74. *FDP* (February 25, 1853).

75. Bolton, *Fugitivism*, 107.

76. Laroche, "Underground Railroad," 108.

77. Walker, "Sailing to Freedom," 30.

78. Susanna Ashton, ed., *I Belong to South Carolina: South Carolina Slave Narratives* (Columbia: University of South Carolina Press, 2010), 75–78, 81 (quotation on 75); Quintana, "Planters, Planters, and Slaves," 116.

79. New York *Evening Post* (September 5, 1853).

80. Philadelphia *North American* (September 13, 1855).

81. *Trenton* (New Jersey) *State Gazette* (March 25, 1854). For another account of this escape, see *Boston Daily Atlas* (March 24, 1854).

82. *NE* (July 5, 1855).

83. Blackett, *Captive's Quest for Freedom*, 243–44.

84. Churchill, *Underground Railroad*, 127 (quotation), 153 (quotation), 10, 156, 207 (quotation).

85. *CA* (August 26, 1837).

86. *Albany* (New York) *Evening Journal* (October 8, 1850). For accounts of other arrivals in Canada via steam, see *CA* (July 21, 1838) and Boston *Emancipator* (November 18, 1846).

87. *FDP* (December 3, 1852).

88. Gilbert H. Barnes and Dwight L. Dumond, eds., *Letters of Theodore Dwight Weld, Angelina Grimké Weld, and Sarah Grimké, 1822–1844*, vol. 2 (1934; repr., New York: Da Capo, 1970), 735.

89. *CA* (July 21, 1838).

90. *FDP* (October 2, 1851).

91. *CA* (October 2, 1841).

92. *NE* (December 30, 1852).

93. *NE* (February 10, 1853).

94. *NE* (September 6, 1855).

95. Foner, *Gateway to Freedom*, 21–22.

96. *FDP* (June 10, 1853).

97. *FDP* (November 17, 1854). For additional accounts mimicking corporate reporting or emphasizing the railroad metaphor, see *FDP* (December 25, 1851; August 12, 1853; September 8, 1854; January 26, 1855; and December 14, 1855) and *Portage County* (Ravenna, OH) *Democrat* (October 11, 1854).

98. *CA* (September 26, 1840).

99. *CA* (December 25, 1841).

100. Sydney Howard Gay, "Record of Fugitives," Columbia University Libraries, https://exhibitions.library.columbia.edu/exhibits/show/fugitives/record_fugitives/transcription, 3, 4, 11, 12, 13, 18, 21. For more on this remarkable document, see Foner, *Gateway to Freedom*, chap. 6.

101. Bolton, *Fugitivism*, 66.

102. For the provisions of the act, see Churchill, *Underground Railroad*, 139–41.

103. *Liberator* (December 24, 1836).

104. Cincinnati *Philanthropist* (September 29, 1837).

105. Casenovia (New York) *Union Herald* (May 18, 1838).

106. *FDP* (November 16, 1849).

107. *FDP* (November 12, 1852).

108. Baltimore *Sun* (September 26, 1854).

109. *Milwaukee Sentinel* (March 20, 1854) and Philadelphia *Public Ledger* (March 18, 1854).

110. Churchill, *Underground Railroad*, 220–22 (quotations on 222).

111. Churchill, *Underground Railroad*, 195, 227–28.

112. Jon Sterngass, "African American Workers and Southern Visitors at Antebellum Saratoga Springs," *American Nineteenth Century History* 2 (2001), 45.

113. Wong, *Neither Fugitive nor Free*, 9, 104.

114. *Alexandria* (Virginia) *Gazette* (July 17, 1847).

115. Boston *Emancipator* (July 21, 1847).

116. Leslie Wheeler, ed., *Loving Warriors: Selected Letters of Lucy Stone and Henry B. Blackwell, 1853 to 1893* (New York: Dial, 1981), 93–4. See also the newspaper accounts in *FDP* (September 8 and 22, 1854), Hartford *Connecticut Courant* (September 9, 1854), and *Boston Herald* (September 1, 1854); and Blackett, *Captive's Quest for Freedom*, 265–66.

117. Schenectady (New York) *Cabinet* (May 30, 1854); for additional accounts of this incident, see *Boston Evening Transcript* (May 27, 1854) and *FDP* (June 2, 1854).

118. *FDP* (March 23, 1855). There were several subsequent legal battles. See Churchill, *Underground Railroad*, 215; and Blackett, *Captive's Quest for Freedom*, 247–48.

119. Richard Archer, *Jim Crow North: The Struggle for Equal Rights in Antebellum New England* (New York: Oxford University Press, 2017), chap. 7.

120. Pryor, *Colored Travelers*, 1, 45.

121. Worcester *Massachusetts Spy* (October 20, 1841).

122. *Douglass' Monthly* (April 1859). For an examination of discrimination faced by Black passengers on streetcars in the antebellum era, see Kelley, *Right to Ride*.

123. Charles B. George, *Forty Years on the Rail*, 3d ed. (Chicago: R. R. Donnelley and Sons, 1887), 221.

124. *CA* (November 18, 1837). "This city" is a reference to New York City.

125. *CA* (November 25, 1837).

126. *CA* (September 30, 1837).

127. Frankie Hutton, *The Early Black Press in America, 1827 to 1860* (Westport, CT: Greenwood, 1993), 82, 103; and Pryor, *Colored Travelers*, 69, 88.

128. *CA* (June 30, 1838).

129. *CA* (August 22, 1840). For other accounts of discrimination, see *CA* (September 1, 1838; December 1, 1838; and July 27, 1839).

130. *CA* (September 2, 1837).

131. *CA* (August 25, 1838).

132. Hodges, *David Ruggles*, 164–66.

133. *NS* (April 14, 1848).

134. *NS* (July 7, 1848).

135. *FDP* (August 20, 1852).

136. *CA* (July 20, 1839).

137. *CA* (September 19, 1840). For other accounts of discrimination, see *CA* (June 27 and July 11, 1840).

138. Entry for August 24, 1836, Connecticut to Philadelphia travel diary (Am.3825), HSP.

139. *CA* (September 25, 1841).

140. Pryor, *Colored Travelers*, 83.

141. *NE* (September 28, 1854).

142. *FDP* (March 23, 1855).

143. *CA* (June 16, 1838).

Chapter 6 • White Women Passengers

1. Sara Chester Smith to Erastus Smith, May 6, 1806, CL 515, Clark Papers, HL.

2. Mary Beth Norton, *Separated by Their Sex: Women in Public and Private in the Colonial Atlantic World* (Ithaca, NY: Cornell University Press, 2011), xii.

3. Norton, *Separated by Their Sex*, xiv (quotation), 152.

4. Norton, *Separated by Their Sex*, xii, xiv; Mary P. Ryan, *Women in Public: Between Banners and Ballots, 1825–1860* (Baltimore: Johns Hopkins University Press, 1990), 37 (quotation), 67. For physical space generally, see Linda K. Kerber, "Separate Spheres, Female Worlds, Woman's Place: The Rhetoric of Women's History," *Journal of American History* 75 (1988), 31.

5. Abigail A. van Slyck, "The Lady and the Library Loafer: Gender and Public Space in Victorian America," *Winterthur Portfolio* 31 (1996), 224.

6. Amy G. Richter, *Home on the Rails: Women, the Railroad, and the Rise of Public Domesticity* (Chapel Hill: University of North Carolina Press, 2005), 5, 33–34.

7. Asa Sheldon, *Yankee Drover: Being the Unpretending Life of Asa Sheldon, Farmer, Trader, and Working Man, 1788–1870* (1862; repr., Hanover, NH: University Press of New England, 1988), 190. For a southern view of this issue, see Marise Bachand, "Gendered Mobility and the Geography of Respectability in Charleston and New Orleans, 1790–1861," *Journal of Southern History* 81 (2015), 42.

8. Unknown to Ruth Tobin, March 27, 1815, Lydia Hollingsworth Correspondence, Maryland Center for History and Culture, Baltimore.

9. Mary Kelley, *Learning to Stand and Speak: Women, Education, and Public Life in America's Republic* (Chapel Hill: University of North Carolina Press, 2006), 78 (quotation), 80; Lucia McMahon, *Mere Equals: The Paradox of Educated Women in the Early Republic* (Ithaca, NY: Cornell University Press, 2012), 140.

10. John H. White, Jr., *Wet Britches and Muddy Boots: A History of Travel in Victorian America* (Bloomington: Indiana University Press, 2013), 448.

11. Directors meeting of January 17, 1837, minutebook of the directors and stockholders (1835–1837), Boston and Worcester Railroad, vol. 2, BARC.

12. Minutes of the meeting of the board of directors, May 14, 1840, vol. 2, Western Railroad, BL.

13. *Yankee Farmer* (May 1840), 166.

14. Quoted in Charles W. Turner, *Chessie's Road* (Richmond, VA: Garrett and Massie, 1956), 33.

15. Quoted in Katherine C. Grier, *Culture and Comfort: Parlor Making and Middle-class Identity, 1850–1930* (Washington, DC: Smithsonian Institution Press, 1997), 47.

16. Charles P. Dare, *Philadelphia, Wilmington and Baltimore Railroad Guide: Containing a Description of the Scenery, Rivers, Towns, Villages, and Objects of Interest along the Line of Road: Including Historical Sketches, Legends, &c.* (Philadelphia: Fitzgibbon & Van Ness, 1856), 71.

17. Richter, *Home on the Rails*, 94–95.

18. Walter Johnson, *River of Dark Dreams: Slavery and Empire in the Cotton Kingdom* (Cambridge, MA: Belknap Press of Harvard University Press, 2013), 131.

19. Ryan, *Women in Public*, 78.

20. Barbara Penner, "'Colleges for the Teaching of Extravagance': New York Palace Hotels," *Winterthur Portfolio* 44 (2010), 159 (quotation), 173; A. K. Sandoval-Strausz, *Hotel: An American History* (New Haven, CT: Yale University Press, 2007), 35–36.

21. Robert Gudmestad, *Steamboats and the Rise of the Cotton Kingdom* (Baton Rouge: Louisiana State University Press, 2011), 59, 75 (quotation).

22. Elisabeth Koren, *The Diary of Elisabeth Koren*, ed. David T. Nelson (Northfield, MN: Norwegian-American Historical Association, 1955), 73 (quotation), 74.

23. John F. Kasson, *Rudeness and Civility: Manners in Nineteenth-Century Urban America* (New York: Hill and Wang, 1990), 132.

24. C. Dallett Hemphill, *Bowing to Necessities: A History of Manners in America, 1620–1860* (New York: Oxford University Press, 1999), 192, 195.

25. Patricia Cline Cohen, "Safety and Danger: Women on American Public Transport, 1750–1850," in *Gendered Domains: Rethinking Public and Private in Women's History*, ed. Dorothy O. Helly and Susan M. Reverby (Ithaca, NY: Cornell University Press, 1992), 122.

26. Cohen, "Safety and Danger," 119.

27. Warren Burton, "Moral Dangers of the City," 1848, AAS-B.

28. *Tricks and Traps of Seducers* (New York: Dinsmore & Co., 1858), 44, 45.

29. "The 'Unprotected Female,'" *GLB* (January 1853).

30. Miss Leslie, *The Behaviour Book: A Manual for Ladies* (Philadelphia: Willis P. Hazard, 1853), 100.

31. Florence Hartley, *The Ladies' Book of Etiquette, and Manual of Politeness* (Philadelphia: G. G. Evans, 1860), 34.

32. *Rules and Regulations to Be Observed by the Men in the Employment of the Mich. Central Railroad Company. Detroit, June, 1850* (Detroit: Duncklee, Wales, & Co., 1850), 4, 10, AAS.

33. *Eastern Rail Road and Branches. Winter Arrangement. Rules for Running Trains, &c., Commencing December 3, 1855* (Salem, MA: William Ives & Co., 1855), 11.

34. Lemuel Porter, *Friendly Advice to Railroad Men. A Discourse Occasioned by the Death of Mr. Charles W. Nichols, an Engineer on the Western Railroad. Delivered in the*

Baptist Church, Pittsfield, Mass., September 7, 1856, by Rev. Lemuel Porter, D. D. (Springfield, MA: Samuel Bowles & Company, 1856), 7, AAS.

35. Kathryn Carlisle Schwartz, *Baptist Faith in Action: The Private Writings of Maria Baker Taylor, 1813–1895* (Columbia: University of South Carolina Press, 2003), 22. The letters are from August 8 and 28, 1833.

36. Carol Bleser, ed., *Tokens of Affection: The Letters of a Planter's Daughter in the Old South* (Athens: University of Georgia Press, 1996), 339–40. The letter is from April 30, 1842.

37. *NE* (August 12, 1847).

38. Edward Hitchcock, *The Power of Christian Benevolence Illustrated in the Life and Labors of Mary Lyon*, 3d ed. (Northampton, MA: Hopkins, Bridgman, and Company, 1852), 97, 244.

39. Sarah Bell Langhorne Waller to Henry Waller, November 25, 1848, box 4(1848), Waller Papers, HL.

40. Lucretia Mott, *Selected Letters of Lucretia Coffin Mott*, ed. Beverly Wilson Palmer (Urbana: University of Illinois Press, 2002), 61.

41. Entries for July 20, August 10, and August 11, 1854, journal 5, Caroline Barrett White Journals, AAS.

42. Ann D. Gordon, ed., *The Selected Papers of Elizabeth Cady Stanton and Susan B. Anthony*, vol. 1, *In the School of Anti-Slavery, 1840 to 1866* (New Brunswick, NJ: Rutgers University Press, 1997), 308.

43. Hemphill, *Bowing to Necessities*, 192.

44. Richter, *Home on the Rails*, 50.

45. March 28, 1831, Rebecca Russell Lowell Gardner travel diary, MHS.

46. *Etiquette for Ladies; with Hints on the Preservation, Improvement, and Display of Female Beauty* (Philadelphia: Carey, Lea & Blanchard, 1838), 11, 12. For another example of advice on acquaintanceships while traveling, see "An English Lady of Rank," *The Ladies' Science of Etiquette* (New York: Wilson & Company, 1844), 4, AAS.

47. A Lady of New York, *Etiquette for Ladies; A Manual of the Most Approved Rules of Conduct in Polished Society, for Married and Unmarried Ladies* (New York: Burgess, Stringer, and Co., 1844), 12, AAS.

48. Leslie, *Behaviour Book*, 100.

49. Richard Lyman Bushman, *The Refinement of America: Persons, Houses, Cities* (New York: Knopf, 1992), chap. 2.

50. Cohen, "Safety and Danger," 119.

51. Kasson, *Rudeness and Civility*, 52.

52. Hemphill, *Bowing to Necessities*, 78.

53. Mme. Celnart, *The Gentleman and Lady's Book of Politeness and Propriety of Deportment, Dedicated to the Youth of Both Sexes*, 2d American ed. (Boston: Allen and Ticknor, 1833), 159, AAS.

54. Samuel Roberts Wells, *How to Behave. A Pocket Manual of Republican Etiquette, and Guide to Correct Personal Habits, Embracing an Exposition of the Principles of Good Manners; Useful Hints on the Care of the Person, Eating, Drinking, Exercise, Habits, Dress, Self-Culture, and Behavior at Home; the Etiquette of Salutations, Introductions, Receptions, Visits, Dinners, Evening Parties, Conversation, Letters, Presents, Weddings, Funerals, the Street, the Church, Places of Amusement, Traveling, etc., with Illustrative Anecdotes, a*

Chapter on Love and Courtship, and Rules of Order for Debating Societies (New York: Fowler & Wells, 1856), 107, AAS.

55. Mrs. John Farrar, *The Young Lady's Friend. By a Lady*, first ed. (Boston: American Stationers' Company, 1836), 406, AAS.

56. *True Politeness: or, Etiquette for Ladies and Gentlemen. Containing the Rules and Usages of Polite Society, with Hints on Courtship and Matrimony. Also, Directions for the Toilette. Compiled from the Most Authentic Sources* (Boston: n.p., 1846), 41, AAS.

57. Farrar, *Young Lady's Friend* (1836), 402–3.

58. *Means and Ends, or Self-Training*, 2d ed. (New York: Harper & Brother, 1842), 153, 156.

59. Miss Sedgwick, *Morals of Manners; or, Hints for our Young People* (New York: Wiley and Putnam, 1846), 20, AAS.

60. Mrs. L. G. Abell, *Woman in Her Various Relations: Containing Practical Rules for American Females, the Best Methods for Dinners and Social Parties—A Chapter for Young Ladies, Mothers, and Invalids—Hints on the Body, Mind, and Character—with a Glance at Woman's Rights and Wrongs, Professions, Costume, etc. etc.* (New York: William Holdredge, 1851), 151.

61. Wells, *How to Behave*, 108.

62. Thomas Mooney, *Nine Years in America*, 2d ed. (Dublin: James McGlashan, 1850), 17, HL.

63. George Templeton Strong, *Diary*, vol. 2, *The Turbulent Fifties, 1850–1859*, ed. Allan Nevins and Milton Halsey Thomas (New York: Macmillan, 1952), 130.

64. Mrs. John Farrar, *The Young Lady's Friend*, rev. ed. (New York: Samuel S. & William Wood, 1849), 368–70.

65. Farrar, *Young Lady's Friend* (1836), 404.

66. Farrar, *Young Lady's Friend* (1836), 413.

67. Abell, *Woman in Her Various Relations*, 150. For drawing as an aspect of women's education, see Kelley, *Learning to Stand and Speak*, 69, 100.

68. *Means and Ends*, 147.

69. Laura F. Edwards, *Only the Clothes on Her Back: Clothing and the Hidden History of Power in the Nineteenth-Century United States* (New York: Oxford University Press, 2022), 115.

70. Farrar, *Young Lady's Friend* (1836), 405.

71. *Means and Ends*, 190.

72. Hartley, *Ladies' Book of Etiquette*, 35.

73. Leslie, *Behaviour Book*, 93, 96.

74. Lincoln Clark to Julia Annah Clark, January 12, 1853, CL 341, Clark Papers, HL.

75. Lincoln Clark to Julia Annah Clark, September 16, 1854, CL 390, Clark Papers, HL.

76. James P. McClure, Peg A. Lamphier, and Erika M. Kreger, eds., *"Spur Up Your Pegasus": Family Letters of Salmon, Kate, and Nettie Chase, 1844–1873* (Kent, OH: Kent State University Press, 2009), 153.

77. Letter from Lucian Barbour, December 19, 1859, folder labeled "Correspondence 1859," box 1, Lucian Barbour Papers, LoC.

78. *GLB* (June 1842), 297.

79. Emma Embury, "Giles Grimstone, The Miser," *GLB* (July 1841), 25.

80. Miss C. M. Sedgwick, "A Day in a Railroad Car," *GLB* (July 1842), 51, 52.
81. Mrs. A. M. F. Annan, "The Midsummer Guests," *GLB* (July 1845), 18.
82. Mrs. S. C. Hall, "Gossip Stings," *GLB* (July 1846), 24.
83. Joseph Chandler, "Gertrude; Or, the Fatal Prophecy," *GLB* (May 1843), 233.
84. *LR* (April 1847), 105.
85. David M. Stone, "The Ring," *GLB* (April 1849), 258.
86. *NE* (September 1, 1859). For another woman contemptuous of her "protector," see Cohen, "Safety and Danger," 122.

Chapter 7 • Children

1. H. D. Barnard, *Travels by Land and Water* (Hartford, CT: H. D. Barnard, 1860), 11, 15–16, 21–22, 29, AAS. The pamphlet states that Barnard was born March 23, 1849, but "1849" is crossed out and "1851" written in by hand, so I take him to be nine years old at the time of publication. For the use of amateur presses among American youth, see Elizabeth Harris, *The Boy and His Press* (Washington, DC: Smithsonian Institution, 1992).

2. Gail Schmunk Murray, *American Children's Literature and the Construction of Childhood* (New York: Twayne, 1998), xv (quotation) and chap. 2; Karen Sanchez-Eppler, *Dependent States: The Child's Part in Nineteenth-Century American Culture* (Chicago: University of Chicago Press, 2005), 7.

3. R. Gordon Kelly, "Literature and the Historian," in *Locating American Studies: The Evolution of a Discipline*, ed. Lucy Maddox (Baltimore: Johns Hopkins University Press, 1999), 104–5. Sarah Maza refers to this technique as telling history "through" children. Sarah Maza, "The Kids Aren't All Right: Historians and the Problem of Childhood," *American Historical Review* 125 (2020), 1261–85.

4. Steven Mintz, *Huck's Raft: A History of American Childhood* (Cambridge, MA: Belknap Press of Harvard University Press, 2004), 76 (quotations); Joseph Illick, *American Childhoods* (Philadelphia: University of Pennsylvania Press, 2002), chaps. 4 and 5; Harvey J. Graff, *Conflicting Paths: Growing Up in America* (Cambridge, MA: Harvard University Press, 1995), chap. 3; and Joseph Kett, "Adolescence and Youth in Nineteenth-Century America," *Journal of Interdisciplinary History* 2 (1971), 283–98.

5. Courtney Weikle-Mills, *Imaginary Citizens: Child Readers and the Limits of American Independence, 1640–1868* (Baltimore: Johns Hopkins University Press, 2013), 103–4.

6. Elizabeth I. Speed to John James Speed, May 6, 1823, HM61173, Speed Family Papers, HL.

7. John D. Long, *The Journal of John D. Long*, ed. Margaret Long (Rindge, NH: Richard R. Smith, 1956), 7, 9, 11, 16.

8. Samuel Thorne, *The Journal of a Boy's Trip on Horseback, Kept by Samuel Thorne* (New York: Privately printed, 1936), 30.

9. Inez McClintock and Marshall McClintock, *Toys in America* (Washington, DC: Public Affairs, 1961), 81–82, 89.

10. Pierce Carlson, *Toy Trains: A History* (New York: Harper and Row, 1986), 15–16; Richard O'Brien, *The Story of American Toys: From the Puritans to the Present* (New York: Abbeville Press, 1990), 19 (quotation).

11. Robert K. Weis, "'To Please and Instruct the Children,'" *Essex Institute Historical Collections* 123 (1987), 122.

12. Alfred Rix and Chastina W. Rix, *New England to Gold Rush California: The Journal of Alfred and Chastina W. Rix, 1849–1854*, ed. Lynn A. Bonfield (Norman, OK: Arthur H. Clark, 2011), 279.

13. Pittsfield Commercial & Classical Boarding School, "Pittsfield Commercial & Classical Boarding School," 1838, AAS-B.

14. "The Subscribers, Principals of the Quaboag Seminary, Warren, Mass.," between 1845 and 1848, lithograph, AAS.

15. St. Mark's Hall, "St. Mark's Hall," 1846; Rectory School (Hamden, CT), "Rectory School," 1851; Raymond Collegiate Institute, "Raymond Collegiate Institute, for young ladies," 1852; Saxton's River Seminary, "Saxton's River Seminary. The spring term of this institution . . . ," 1853; and Boscawen Academy, "Family and day school in the country," 1857, all in AAS-B.

16. Lincoln Clark to Julia Annah Smith Clark, January 9, 1852, CL 259, Clark Papers, HL.

17. Julia Annah Clark to Catherine Lincoln Clark, September 23, 1859, CL 554, Clark Papers, HL.

18. *NE* (October 6, 1859).

19. *FDP* (November 24, 1854).

20. *Baltimore Gazette* (October 25, 1832).

21. *Lafayette* (Indiana) *Daily Journal* (April 25, 1854).

22. Ruth Miller Elson, *Guardians of Tradition: American Schoolbooks of the Nineteenth Century* (Lincoln: University of Nebraska Press, 1964), 1, 245 (quotation).

23. John B. Longgley, *The Youth's Companion, or a Historical Dictionary; Consisting of Articles Selected Chiefly from Natural and Civil History, Geography, Astronomy, Zoology, Botany and Mineralogy* (St. Clairsville, OH: Printed for the Compiler by Horton J. Howard, 1832), 55–56, 268–69 (quotation).

24. Lyman Cobb, *The North American Reader; Containing a Great Variety of Pieces in Prose and Poetry, from Very Highly Esteemed American and English Writers* [. . .] (New York: Harper & Brothers, 1835), 412–14 (quotation on 413).

25. P. H. Snow, *The American Reader: Containing Selections in Prose, Poetry and Dialogue. Designed for the Use of Advanced Classes in Public Schools, High Schools, and Academies* (Hartford, CT: Spalding and Storrs, 1840), 127–29, AAS. For the characterization of silence as unproductive, see Mark M. Smith, *Listening to Nineteenth-Century America* (Chapel Hill: University of North Carolina Press, 2001), 163.

26. Snow, *American Reader*, 156–57. For the treatment of Native Americans in school literature, see Elson, *Guardians of Tradition*, 79; and Carolyn Eastman, "The Indian Censures the White Man: 'Indian Eloquence' and American Reading Audiences in the Early Republic," *William and Mary Quarterly*, 3d ser., 65 (2008), 538.

27. Charles W. Sanders, *The School Reader. Third Book* (New York: Mark H. Newman, 1848), 121–22.

28. *The Little Keepsake: A Poetic Gift for Children* (New York: Kiggins and Kellogg, 1849), AAS.

29. John Lauritz Larson, *Internal Improvement: National Public Works and the Promise of Popular Government in the Early United States* (Chapel Hill: University of North Carolina Press, 2001), 235.

30. Elson, *Guardians of Tradition*, 260.

31. Howard P. Chudacoff, *How Old Are You? Age Consciousness in American Culture* (Princeton, NJ: Princeton University Press, 1989), 21.

32. Daniel T. Rodgers, "Socializing Middle-Class Children: Institutions, Fables, and Work Values in Nineteenth-Century America," *Journal of Social History* 13 (1980), 357.

33. Ronald J. Zboray, *A Fictive People: Antebellum Economic Development and the American Reading Public* (New York: Oxford University Press, 1993), xvii.

34. William Darby, *The United States Reader, or Juvenile Instructor, No. 2* (Baltimore: Plaskitt & Co., 1829), 133–36. I am indebted to David Schley for providing this transcription.

35. Walter Aimwell, *Whistler; Or, the Manly Boy* (Boston: Gould and Lincoln, 1857), 17–18, AAS.

36. Aimwell, *Whistler*, 22–23.

37. Jacob Abbott, *Rollo's Vacation* (Boston: Phillips, Sampson, and Company, 1855), 118–19, AAS. The book was originally published much earlier; the "Prefatory Notice" at the beginning of the book is dated October 1, 1838.

38. Elisha Noyce, *The Boy's Book of Industrial Information* (New York: D. Appleton & Co., 1859), AAS.

39. John Comly, *Comly's Reader, and Book of Knowledge; with Exercises of Spelling and Defining, Intended for the Use of Schools, and for Private Instruction* (Philadelphia: Thomas L. Bonsal, 1845), 145.

40. Noah Webster, *The Pictorial Elementary Spelling Book; Being an Improvement on the American Spelling Book* (New York: George F. Cooledge & Brother, 1844), 54, AAS.

41. *The National Pictorial Primer; Designed for the Use of Schools and Families, Embellished with More than One Hundred and Fifty Fine Engravings* (New York: George F. Cooledge and Brother, 1846), 29, 46.

42. B. D. Emerson, *The Second-Class Reader: Designed for the Use of the Middle Class of Schools in the United States* (Philadelphia: Hogan and Thompson, 1841), 165–68. For a general discussion of these themes, see Murray, *American Children's Literature*, chap. 2.

43. T. S. Pinneo, *The Hemans Reader for Female Schools: Containing Extracts in Prose and Poetry, Selected from the Writings of More than One Hundred and Thirty Different Authors* (New York: Pratt, Woodford and Company, 1847), 338–40.

44. Sanders, *School Reader*, 80–81, 83.

45. *The Skating Party, and Other Stories* (New York: Leavitt & Allen, 1855), 8–10, AAS.

46. Comly, *Comly's Reader*, 192.

47. Author of Four Days in July, *What Norman Saw in the West* (New York: Carlton & Porter for the Sunday-School Union, 1859), 268, AAS.

48. Katherine Pandora, "The Children's Republic of Science in the Antebellum Literature of Samuel Griswold Goodrich and Jacob Abbott," *Osiris* 24 (2009), 78.

49. Jacob Abbott, *Aunt Margaret; Or, How John True Kept His Resolutions* (New York: Harper & Brothers, 1856), 14, 16, AAS.

50. Abbott, *Aunt Margaret*, 17; Scott Sandage, *Born Losers: A History of Failure in America* (Cambridge, MA: Harvard University Press, 2005), 193. The notion of dependency was deep-seated in both the North and the South, even as the two regions had different realities of "dependency." See Mary Ryan, *Cradle of the Middle Class:*

The Family in Oneida County, New York, 1790–1985 (New York: Cambridge University Press, 1981), chap. 4; Sean Wilentz, *Chants Democratic: New York City and the Rise of the American Working Class, 1788–1850* (New York: Oxford University Press, 1984); Lacy K. Ford, Jr., *Origins of Southern Radicalism: The South Carolina Upcountry, 1800–1860* (New York: Oxford University Press, 1988), 351; Stephanie McCurry, *Masters of Small Worlds: Yeoman Households, Gender Relations, and the Political Culture of the Antebellum South Carolina Lowcountry* (New York: Oxford University Press, 1995); and Amy Dru Stanley, *From Bondage to Contract: Wage Labor, Marriage, and the Market in the Age of Slave Emancipation* (New York: Cambridge University Press, 1998). For fear and child-raising, see Peter N. Stearns and Timothy Haggerty, "The Role of Fear: Transitions in American Emotional Standards for Children, 1850–1950," *American Historical Review* 96 (1991), 64–94; and Rodgers, "Socializing Middle-Class Children," 358.

51. Patricia Crain, *Reading Children: Literacy, Property, and the Dilemmas of Childhood in Nineteenth-Century America* (Philadelphia: University of Pennsylvania Press, 2016), 102.

52. Abbott, *Aunt Margaret*, 52, 55, 61–66, 68, 84.

53. Jacob Abbott, *The Three Gold Dollars; Or, An Account of the Adventures of Robin Green* (New York: Harper and Brothers, 1856), 62–63, 75–77, 80–85, AAS. Travel also figured prominently in another one of Abbott's novels: Jacob Abbott, *The Great Elm; Or, Robin Green and Josiah Lane at School* (New York: Harper and Brother, 1856), AAS.

54. Henry Webster Parker, *A Summer with the Little Grays* (Boston: Walker, Wise, & Co., 1860), 122–25, 128–29, AAS. Orphans hold a significant place in nineteenth-century literature, and the children's literature considered here was no exception. Mintz, *Huck's Raft*, 157.

55. Harley Thorne, ed., *Youth's Casket; an Illustrated Magazine for Children*, vol. 1 (Buffalo, NY: Beadle and Brother, 1852), 129.

Conclusion

1. Michael C. Cohen, *The Social Lives of Poems in Nineteenth-Century America* (Philadelphia: University of Pennsylvania Press, 2015), 23.

ESSAY ON SOURCES

Primary Sources

In writing this book, I used an eclectic set of primary sources to find references to steam transit in a wide variety of times and locations. I made use of numerous printed and manuscript diaries and letter collections, which is clear from the notes. For manuscripts, I consulted materials at the American Antiquarian Society, the Baker Library at Harvard Business School, the Hargrett Library at the University of Georgia, the Historical Society of Pennsylvania, the Huntington Library, the Library Company of Philadelphia, the Library of Congress, the Maryland Center for History and Culture, the Massachusetts Historical Society, the South Caroliniana Library at the University of South Carolina, the Virginia Historical Society, and the Wisconsin Historical Society.

Beyond the diaries and letter collections, I also made use of a few specific types of printed materials. The printed annual reports of transportation companies have often been mined for their relevance to political and economic history, but there is a great deal of social history in these materials as well. Comic almanacs gave me access to many jokes and cartoons about transportation. Etiquette guides gave Americans standards on how to behave on these new forms of transit. Printed sermons and eulogies gave ministers an opportunity to share their own thoughts about steam transit. The Black press was a major source for learning about how enslaved people used steam transit for their escapes, as well as the treatment of Black Americans in the North—*Colored American*, *Frederick Douglass' Paper*, *National Era*, and *North Star* were my main sources here. For literature aimed at women, *Ladies' Repository* and *Godey's Lady's Book* were the periodicals I used the most. Children's literature, particularly textbooks used in schools and novels, were instrumental in my understanding of how young Americans encountered these machines. Guidebooks gave detailed descriptions of their different routes. Sheet music captured the efforts of composers to honor or imitate steam transit in their works. Finally, the American Antiquarian Society has a massive collection of broadsides which was integral to my research.

Secondary Sources

In my work I have benefited from a wide array of secondary literature. The Transportation Revolution has scholarship stretching back several decades. A critical early work is George Rogers Taylor's *The Transportation Revolution, 1815–1860* (New York: Rinehart, 1951). Albert Fishlow's comprehensive essay, "Internal Transportation in the Nineteenth and Early Twentieth Centuries," provides a critical overview of the economic fruits of this revolution (see *The Cambridge Economic History of the United States*, vol. 2, *The Long Nineteenth Century*, edited by Stanley L. Engerman and Robert E. Gallman [New York: Cambridge University Press, 2000], 543–642). The works of John Lauritz Larson, including *Internal Improvement: National Public Works and the Promise of Popular Government in the Early United States* (Chapel Hill: University of North Carolina Press, 2001) and *Market Revolution in America: Liberty, Ambition, and the Eclipse of the Common Good* (New York: Cambridge University Press, 2010), have been critical to my understanding of the larger period of American history. Daniel Walker Howe's *What Hath God Wrought: The Transformation of America, 1815–1848* (New York: Oxford University Press, 2007) provides an exhaustive overview of the era.

More recently, scholars have worked to provide a more nuanced understanding of the Transportation Revolution, including its social and cultural aspects. One key early work here is Richard H. Gassan's *Birth of American Tourism: New York, the Hudson Valley, and American Culture, 1790–1830* (Amherst: University of Massachusetts Press, 2008). William Mackintosh's *Selling the Sights: The Invention of the Tourist in American Culture* (New York: New York University Press, 2019) helped me understand of the commodification of travel. Robert Gudmestad's *Steamboats and the Rise of the Cotton Kingdom* (Baton Rouge: Louisiana State University Press, 2011) is a compelling and wide-ranging history of steamboats on the western waters that provided a sorely needed update on this important topic. John H. White, Jr.'s *Wet Britches and Muddy Boots: A History of Travel in Victorian America* (Bloomington: Indiana University Press, 2013) gives an overview of the travel experience in the nineteenth century. A. K. Sandoval-Strausz's *Hotel: An American History* (New Haven, CT: Yale University Press, 2007) is an important history of an institution that helped make travel possible. Elizabeth Stordeur Pryor's *Colored Travelers: Mobility and the Fight for Citizenship before the Civil War* (Chapel Hill: University of North Carolina Press, 2016) explores the history of Black mobility. David Schley's *Steam City: Railroads, Urban Space, and Corporate Capitalism in Nineteenth-Century Baltimore* (Chicago: University of Chicago Press, 2020) ties transportation history to urban history, and Benjamin Sidney Michael Schwantes's *The Train and the Telegraph: A Revisionist History* (Baltimore: Johns Hopkins University Press, 2019) makes a nuanced argument about the relationship about those two particular technologies. There are numerous histories of individual corporations during this period; two that I found particularly helpful are James D. Dilts's *The Great Road: The Building of the Baltimore & Ohio, the Nation's First Railroad, 1828–1853* (Stanford, CA: Stanford University Press, 1993) and Albert Churella's *The Pennsylvania Railroad*, vol. 1, *Building an Empire, 1846–1917* (Philadelphia: University of Pennsylvania Press, 2013). Sensory history has also been critical to my understanding of the Transportation Revolution; an overview of this growing literature can be found in Mark M. Smith, *A Sensory History Manifesto* (University Park: Pennsylvania State University Press, 2021).

Widening the lens to from the Transportation Revolution to transportation and culture in general, foundational works have included Leo Marx, *The Machine in the Garden: Technology and the Pastoral Ideal in America*, rev. ed. (New York: Oxford University Press, 2000); John Kasson, *Civilizing the Machine: Technology and Republican Values in America, 1776–1900*, rev. ed. (New York: Hill and Wang, 1999); and Wolfgang Schivelbusch, *The Railway Journey*, Anselm Hollo, trans. (New York: Urizen, 1979). Additional works include Eugene Alvarez, *Travel on Southern Antebellum Railroads, 1828–1860* (University: University of Alabama Press, 1974); James A. Ward, *Railroads and the Character of America, 1820–1887* (Knoxville: University of Tennessee Press, 1986); Maury Klein, *Unfinished Business: The Railroad in American Life* (Hanover, NH: University Press of New England, 1994); Carol Sheriff, *The Artificial River: The Erie Canal and the Paradox of Progress, 1817–1862* (New York: Hill and Wang, 1996); John R. Stilgoe, *Train Time: Railroads and the Imminent Reshaping of the United States Landscape* (Charlottesville: University of Virginia Press, 2007); Craig Miner, *A Most Magnificent Machine: America Adopts the Railroad, 1825–1862* (Lawrence: University Press of Kansas, 2010); H. Roger Grant, *Railroads and the American People* (Bloomington: Indiana University Press, 2012); and Julia H. Lee, *The Racial Railroad* (New York: New York University Press, 2022). Other key works on technology and culture include Wiebe E. Bijker, *Of Bicycles, Bakelites, and Bulbs: Toward a Theory of Sociotechnical Change* (Cambridge, MA: MIT Press, 1995); David E. Nye, *America as Second Creation: Technology and Narratives of New Beginnings* (Cambridge, MA: MIT Press, 2003); and Thomas Hughes, *Human-Built World: How to Think about Technology and Culture* (Chicago: University of Chicago Press, 2004). My own *Railroads in the Old South: Pursuing Progress in a Slave Society* (Baltimore: Johns Hopkins University Press, 2009) attempted to better understand the use of railroads in a slave society. Alexis D. Litvine's "Annihilation of Space: A Bad (Historical) Concept" (*Historical Journal* 65 [2022]: 871–900) is an important article on a much overused trope. Two important works about accidents and their implications are Robert B. Shaw, *A History of Railroad Accidents, Safety Precautions, and Operating Practices* (N.p.: Vail-Ballou, 1978); and Mark Aldrich, *Death Rode the Rails: American Railroad Accidents and Safety, 1828–1965* (Baltimore: Johns Hopkins University Press, 2006).

The literature on the Underground Railroad and Black mobility has blossomed in recent years. Works that helped me make sense of this topic include Edlie L. Wong, *Neither Fugitive nor Free: Atlantic Slavery, Freedom Suits, and the Legal Culture of Travel* (New York: New York University Press, 2009); Blair L. M. Kelley, *Right to Ride: Streetcar Boycotts and African American Citizenship in the Era of* Plessy v. Ferguson (Chapel Hill: University of North Carolina Press, 2010); Graham Russell Gao Hodges, *David Ruggles: A Radical Black Abolitionist and the Underground Railroad in New York City* (Chapel Hill: University of North Carolina Press, 2010); R. J. M. Blackett, *Making Freedom: The Underground Railroad and the Politics of Slavery* (Chapel Hill:

University of North Carolina Press, 2013); Matthew Salaria, *Slavery's Borderland: Freedom and Bondage along the Ohio River* (Philadelphia: University of Pennsylvania Press, 2013); Walter Johnson, *River of Dark Dreams: Slavery and Empire in the Cotton Kingdom* (Cambridge, MA: Belknap Press of Harvard University Press, 2013); Eric Foner, *Gateway to Freedom: The Hidden History of the Underground Railroad* (New York: W. W. Norton, 2015); Angela F. Murphy, *The Jerry Rescue: The Fugitive Slave Law, Northern Rights, and the American Sectional Crisis* (New York: Oxford University Press, 2016); Richard Archer, *Jim Crow North: The Struggle for Equal Rights in Antebellum New England* (New York: Oxford University Press, 2017); R. J. M. Blackett, *The Captive's Quest for Freedom: Fugitive Slaves, the 1850 Fugitive Slave Law, and the Politics of Slavery* (New York: Cambridge University Press, 2018); Paul Finkelman, *Supreme Injustice: Slavery in the Nation's Highest Court* (Cambridge, MA: Harvard University Press, 2018); S. Charles Bolton, *Fugitivism: Escaping Slavery in the Lower Mississippi Valley, 1820–1860* (Fayetteville: University of Arkansas Press, 2019); and Robert H. Churchill, *The Underground Railroad and the Geography of Violence in Antebellum America* (New York: Cambridge University Press, 2020). Timothy D. Walker's edited collection of essays *Sailing to Freedom: Maritime Dimensions of the Underground Railroad* (Amherst: University of Massachusetts Press, 2021) is an important book with a rich array of detail on how steamboats were also part of how enslaved people moved north.

For understanding antebellum music and its relationship to transportation, I relied heavily on Nancy Newman, *Good Music for a Free People: The Germania Musical Society in Nineteenth-Century America* (Rochester, NY: University of Rochester Press, 2010); Brian Roberts, *Blackface Nation: Race, Reform, and Identity in American Popular Music, 1812–1925* (Chicago: University of Chicago Press, 2017); and Billy Coleman, *Harnessing Harmony: Music, Power, and Politics in the United States, 1788–1865* (Chapel Hill: University of North Carolina Press, 2020). For the broader context of the role of privately owned pianos, I relied on Arthur Loesser, *Men, Women, and Pianos: A Social History*, new ed. (New York: Dover, 1990), and James Parakilas's edited volume, *Piano Roles: Three Hundred Years of Life with the Piano* (New Haven, CT: Yale University Press, 1999). Other important works include Charles E. Hamm, *Yesterdays: Popular Song in America* (New York: W. W. Norton, 1979); Norm Cohen, *Long Steel Rail: The Railroad in American Folksong* (Urbana: University of Illinois Press, 1981); and Daniel Cavicchi, *Listening and Longing: Music Lovers in the Age of Barnum* (Middletown, CT: Wesleyan University Press, 2011). For visual representations of trains, Ian Kennedy and Julian Treuherz's edited book *The Railway: Art in the Age of Steam* (New Haven, CT: Yale University Press, 2008) is an overview for the time period. I also benefited from John Francis McDermott, *The Lost Panoramas of the Mississippi* (Chicago: University of Chicago Press, 1958); Karin Hertel McGinnis, "Moving Right Along: Nineteenth Century Panorama Painting in the United States" (Ph.D. diss., University of Minnesota, 1983); and Vanessa Meikle Schulman, *Work Sights: The Visual Culture of Industry in Nineteenth-Century America* (Amherst: University of Massachusetts Press, 2015). For literature and reading, I learned from Ronald J. Zboray, *A Fictive People: Antebellum Economic*

Development and the American Reading Public (New York: Oxford University Press, 1993); Kevin J. Hayes, "Railway Reading" (*Proceedings of the American Antiquarian Society* 106 [1997]: 301–26); and Gillian Silverman, *Bodies and Books: Reading and the Fantasy of Communion in Nineteenth-Century America* (Philadelphia: University of Pennsylvania Press, 2012). Works specifically about humor include Constance Rourke, *American Humor: A Study of the National Character* (1931; repr., New York: New York Review of Books, 2004); Marion Barber Stowell, "Humor in Colonial Almanacs" (*Studies in American Humor* 3 [1976]: 34–47); Cameron C. Nickels, *New England Humor: From the Revolutionary War to the Civil War* (Knoxville: University of Tennessee Press, 1993); and James H. Justus, *Fetching the Old Southwest: Humorous Writing from Longstreet to Twain* (Columbia: University of Missouri Press, 2004). The work of Mike Esbester, particularly "Designing Time: The Design and Use of Nineteenth-Century Transport Timetables" (*Journal of Design History* 22, no. 2 [2009]: 91–113) and "Nineteenth-Century Timetables and the History of Reading" (*Book History* 12 [2009]: 156–85), shows how railroads impacted reading.

Works about the interplay of technology and religion include John Hedley Brooke, *Science and Religion: Some Historical Perspectives* (1991; repr., Cambridge: Cambridge University Press, 2014); Jay Newman, *Religion and Technology: A Study in the Philosophy of Culture* (Westport, CT: Praeger, 1997); Thomas Dixon, Geoffrey Cantor, and Stephen Pumfrey, eds., *Science and Religion: New Historical Perspectives* (New York: Cambridge University Press, 2010); and Donald A. Yerxa, ed., *Religion and Innovation: Antagonists or Partners?* (London: Bloomsbury, 2016). Works about antebellum religion that were useful to me include Robert H. Abzug, *Cosmos Crumbling: American Reform and the Religious Imagination* (New York: Oxford University Press, 1994); David Paul Nord, *Faith in Reading: Religious Publishing and the Birth of Mass Media in America* (New York: Oxford University Press, 2004); John Lardas Modern, *Secularism in Antebellum America, with Reference to Ghosts, Protestant Subcultures, Machines, and Their Metaphors; Featuring Discussions of Mass Media, Moby-Dick, Spirituality, Phrenology, Anthropology, Sing Sing State Penitentiary, and Sex with the New Motive Power* (Chicago: University of Chicago Press, 2011); and Jon Gjerde, *Catholicism and the Shaping of Nineteenth-Century America*, edited by S. Deborah Kang (New York: Cambridge University Press, 2012). Specific works about the Sabbath include Richard R. John, "Taking Sabbatarianism Seriously: The Postal System, the Sabbath, and the Transformation of American Political Culture" (*Journal of the Early Republic* 10 [1990]: 517–67); and Alexis McCrossen, *Holy Day, Holiday: The American Sunday* (Ithaca, NY: Cornell University Press, 2000). Drew Gilpin Faust's *This Republic of Suffering: Death and the American Civil War* (New York: Knopf, 2008) was critical to my understanding of the Good Death.

The literature on women in the antebellum era is vast, but a few particular works guided me through the process of understanding the development of private and public spheres from colonial times through the Civil War: Mary P. Ryan, *Women in Public: Between Banners and Ballots, 1825–1860* (Baltimore: Johns Hopkins University Press, 1990); Abigail A. van Slyck, "The Lady and Library Loafer: Gender and Public Space in Victorian America" (*Winterthur Portfolio* 31 [1996]: 221–42); Mary

Kelley, *Learning to Stand and Speak: Women, Education, and Public Life in America's Republic* (Chapel Hill: University of North Carolina Press, 2006); Mary Beth Norton, *Separated by Their Sex: Women in Public and Private in the Colonial Atlantic World* (Ithaca, NY: Cornell University Press, 2011); and Lucia McMahon, *Mere Equals: The Paradox of Educated Women in the Early Republic* (Ithaca, NY: Cornell University Press, 2012). Critical works about etiquette from the era include John F. Kasson, *Rudeness and Civility: Manners in Nineteenth-Century Urban America* (New York: Hill and Wang, 1990); and C. Dallett Hemphill, *Bowing to Necessities: A History of Manners in America, 1620–1860* (New York: Oxford University Press, 1999). Katherine C. Grier, *Culture and Comfort: Parlor Making and Middle-class Identity, 1850–1930* (Washington, DC: Smithsonian Institution Press, 1997), gives an overview of the material culture of the nineteenth century that is helpful to understanding how corporations accommodated women on steam transport. Patricia Cline Cohen, "Safety and Danger: Women on American Public Transport, 1750–1850," in *Gendered Domains: Rethinking Public and Private in Women's History*, edited by Dorothy O. Helly and Susan M. Reverby (Ithaca, NY: Cornell University Press, 1992), pp. 109–22, outlines the challenges that women faced on transportation. And although Amy Richter's *Home on the Rails: Women, the Railroad, and the Rise of Public Domesticity* (Chapel Hill: University of North Carolina Press, 2005) is principally concerned with the postbellum era, it has been a vital work for my own understanding of how women negotiated steam transit.

Useful literature on children includes Howard P. Chudacoff, *How Old Are You? Age Consciousness in American Culture* (Princeton, NJ: Princeton University Press, 1989); Harvey J. Graff, *Conflicting Paths: Growing Up in America* (Cambridge, MA: Harvard University Press, 1995); Joseph Illick, *American Childhoods* (Philadelphia: University of Pennsylvania Press, 2002); Steven Mintz, *Huck's Raft: A History of American Childhood* (Cambridge, MA: Belknap Press of Harvard University Press, 2004); Karen Sanchez-Eppler, *Dependent States: The Child's Part in Nineteenth-Century American Culture* (Chicago: University of Chicago Press, 2005); Courtney Weikle-Mills, *Imaginary Citizens: Child Readers and the Limits of American Independence, 1640–1868* (Baltimore: Johns Hopkins University Press, 2013); and Sarah Maza, "The Kids Aren't All Right: Historians and the Problem of Childhood" (*American Historical Review* 125 [2020]: 1261–85). Two works specifically on children's literature are Gail Schmunk Murray, *American Children's Literature and the Construction of Childhood* (New York: Twayne, 1998); and Patricia Crain, *Reading Children: Literacy, Property, and the Dilemmas of Childhood in Nineteenth-Century America* (Philadelphia: University of Pennsylvania Press, 2016).

INDEX

Page numbers in *italics* indicate figures.

Abbott, Jacob, 199–200, 203–5
abolitionists, 8, 132, 148, 151, 157, 159–60
accidents, 69–71, 96, 192; and corporate liability, 11, 32–33; in fiction, 201; as lessons for children, 198, 200; potential for, 3, 64–67, 183. *See also* death; Protestants: reaction to accidents
Adams, J. G., 120
Adams, John, 183
Adams, John Quincy, 76
Akron, Ohio, 154
Alabama, 40, 136, 141, 155
Albany (steamboat), 58
Albany and West Stockbridge Railroad, 26
Albany, New York, 44, 47, 83
Aleck (an enslaved man), 143–44
Allen, William, 60
Amelia Court House, Virginia, 16
Androscoggin, Maine, 89
anonymity on steam transit, 55–58
Appleton, William, 33
Archbold, Ann, 66–67
Arkansas, 1, 194
Armistead, Rosette (or Rosetta Armstead), 157
artwork. *See* images
Ashworth, Henry, 135
Athens, Georgia, 62
Atlantic (steamboat), 200
Aurora, New York, 189

baggage, 8, 45–49, 60, 171–74, 179, 181, 185
Baldwin, Christopher, 43

Baltimore, Maryland: as a site of dispute, 28; travel from, 41, 87–88, 143, 145, 151, 153; travel to, 43, 46, 181, 196
Baltimore and Ohio Railroad, 13–15, 86
Baltimore and Susquehanna Railroad, 26
Bangor, Maine, 89
Banister, William, 47
Banvard, John, 96
Barber, D. J., 37
Barbour, Lucian, 181
Barnard, H. D., 187–88, 210
Barnwell District, South Carolina, 20
Bear, John, 34
behavior. *See* etiquette
Bellows, Mr., 33
biblical quotations, 105–6, 109
Black Americans. *See* Black mobility; free Black people
Black mobility, 129–64, 209; and commodified travel, 129, 133, 142, 146, 148–49, 164; regulated, 134–42
Blackstone Canal, 14
Blackwell, Henry, 77, 156
Bliss, Alexander, 41
Boardman, Henry, 127
Boston, Massachusetts, 25, 90–91, 96, 111, 170; freight traffic to, 20–21; railroad jubilee in, 37, 111; travel from, 62; travel to, 80, 84, 111, 145, 163
Boston and Maine Railroad, 22, 24
Boston and Worcester Railroad, 14, 29–33, 46, 52, 118, 168

Bradford, Alexander, 107
Breckinridge, John, 92, *94*
Brewer, D. R., 126
Brown, George, 190
Brown, Henry Box, 143
Brownson, Orestes, 120
Buckfield, Maine, 189
Bucks County, Pennsylvania, 190
Buffalo, New York, 43, 155, 206
Bull, Thomas, 41
Burleigh, Charles C., 156
Burlington, New Jersey, 126
Burr, David Auguste, 59
Burton, Warren, 170
Bushnell, Horace, 105–6

Cambridge, Massachusetts, 90
Camden and Amboy Railroad, 70, 126
Canada, Catholics in, 111–12; enslaved people escape to, 133, 138, 143, 146, 148–49, 151
Carroll, Charles, 15, 86
Cass, Lewis, 56, 92
Chamberlain, Samuel, 61
Charleston, South Carolina, 16, 123, 138, 140, 143, 145
Charleston to Hamburg Line. *See* South Carolina Railroad
Charlotte, North Carolina, 29
Chase, Lurana Y. White, 70
Chase, Salmon, 181
Chesapeake and Delaware Canal, 143
Chester, Pennsylvania, 29
Chicago, Illinois, 43, 49, 52, 203
Child, Lydia Maria, 149
children, 8–9; as authors, 187–88; fiction for, 196–97, 199–207; travel by, 189–96
Cincinnati, Columbus and Cleveland Railroad, 106
Cincinnati, Ohio, 87, 90, 101, 145, 157; as a gateway to freedom, 148; travel to, 46, 87, 143, 154, 173
Clark, Julia, 181
Clark, Lincoln, 19, 181, 191; travel description by, 40, 46, 55, 62, 66, 67–68
Clemson, Anna Calhoun, 50
Cleveland, Ohio, 148
Clinton, De Witt, 193
Cobb, Lyman, 193

Cole, Thomas, 16
Columbia, Pennsylvania, 18
Columbia River, 194
Columbus, Ohio, 154
comic almanacs, 47, 50, 78–82, 109
Concord, New Hampshire, 91
Connecticut, 13, 20, 191
Connecticut River Railroad, 30
Connell, Maria Bryan, 172
Converse, Henrietta, 74
Copp, Joseph, 108
corporation and community interaction, 10, 14, 18, 27–31
Covington, Kentucky, 145
Craft, Ellen and William, 144
Crockett, Davy, 78
Cumberland, Maryland, 46
Cumberland Valley Railroad, 120
Currier, Nathaniel, 69

Dall, Caroline Healey, 37, 76, 90, 92
danger, 7, 63–66, 70–71; and children, 192; fictional descriptions of, 183, 201; humorous descriptions of, 80–82, 85; musical depictions of, 91–92; on railroad tracks, 31; and women, 170–71, 180
Davis, Ned, 147
death: cultural attitudes towards, 63–64, 70; "Good Death," 7, 39, 64, 71, 127–28; on railroads, 28, 33, 39, 65–66, 68, 79; on steamboats, 67, 80–81, 124. *See also* accidents; Protestants: reaction to accidents
decorum. *See* etiquette
depots, 26, *48*, 57, 68, 102, 170, 181, 201; alternate uses of, 11, 31, 34–35, 114; construction of, 24–25, 29; demand for, 24–25; opposition to, 29. *See also* images: of depots; slavery: confrontation at depots
Detroit, Michigan, 149, 187
Dimmick, L. F., 118–19
Dodge, Mary Abigail. *See* Hamilton, Gail
Douglas, Stephen, 92, *94*
Douglass, Frederick: on Black mobility, 129–130, 163; on enslavement, 57, 129, 134, 141–42; on equal treatment, 158, 160; on freedom, 155, 157–58
Dover, New Hampshire, 12
Dubuque, Iowa, 19

Index 257

du Pont, Sophie, 58
Dutt, William, 63

Eastern Railroad, 74, 157, 172
Eastman, Charles, 27
East St. Louis, Illinois, 143
East Tennessee and Virginia Railroad, 135
Eaton, Joseph Dowding Bass, 60
Edgefield District, South Carolina, 20
Eglin, Harriet, 145
Elizabethtown, New Jersey, 116
Elliott, William, 43, 61
Elmira, New York, 74
Emerson, B. D., 199
Emerson, Ralph Waldo, 16
Endicott, Charles, 13
entertainers, 54
Erie Railroad, 87, 100
etiquette, 6–8, 71; enforced by passengers, 52–53; established rules, 39, 50, 58, 97; guides, 4, 8, 52, 75, 115, 175–79; for men, 52, 174; for women, 8, 165–66, 170–72, 174–81, 186
eulogies. *See* funerals
E. W. Stevens (steamboat), 138

fiction, 182–85. *See also* children: fiction for; poetry
Fischer, William, 91
Fisher, Sidney George, 43
Fitchburg Railroad, 23, 35
Florida, 16
Floyd, William, 43
food. *See* taste
Forman, Isaac, 137
Foster, Nathan, 13
Frankfort, Kentucky, 57
Franklin and Bristol Railroad, 21
Franklin, Massachusetts, 203–4
Fredericksburg, Virginia, 143
Fredericktown, Maryland, 46
free Black people, 8, 137, 140, 152–53, 157, 163; interaction with enslaved people, 138, 142, 156–57; state regulation of, 137
freight shipments: complaints about, 27; rules about, 11, 26–27
Fugitive Slave Act, 149–50, 152, 154
Fulton (steamboat), 198

funerals, 104, 123–27, 172, 192
Furman, Maria Dorthea, 172

Gannett, Rev. Dr., 119
Gardner, Rebecca Russell Lowell, 174
Garnet, Henry Highland, 160
Gay, Sidney Howard, 151
George, Charles, 158
Georgia, 20, 85, 140–41, 147
Giles, Charlotte, 145
Gordonsville, Virginia, 61
Gray, J. C., 23
Great Lakes, 19, 203. *See also individual lakes*
Green, Betsey, 152
Green Bay, Wisconsin, 66, 158
Grimké, Angelina, 65
Guignard, J., 87
Guild, William, 101
Gungl, Josef, 90
Gunnison, John Williams, 66

Habersham, Robert, 66
Hamburg and Charleston Railroad. *See* South Carolina Railroad
Hamilton, Gail, 185
Hamlin, Hannibal, 92, *94*
Hammond, Edward Spann, 134
Hanna, Esther Belle, 121
Harbaugh, F. Reck, 126–27
Harlem Railroad, 28, 91
Harrisburg, Pennsylvania, 148, 153
Havre de Grace, Maryland, 45
hearing, 1–2, 67, 75; music, 86–92; sounds of steam transit, 60–61, 97, 107, 156, 197, 206; warning sounds, 32, 94. *See also* danger
Hedges, M. J., 74
Helm, Daniel, 45
Hinckley, Isaac, 38
Hixon, Charlotte, 59, 109
Hoge, Moses, 50
Holmes, Oliver Wendell, 79
Home (steamboat), 123
Hopkins, Mrs. S., 41–42
Horry, Harriott Pinckney, 13, 50
hotels, 40, 71, 169
Hudson, New York, 99
Hudson River, 59, 99
Hudson River Railroad, 43, 47, 100

Hudson Valley, 57, 193
humor, 7, 77–86, 102; dark humor, 72, 79–82, *83*; mocking inexperienced travelers, *48*, 84–86; puns, 79–82, 109; steam transit as the setting, 72, 82, 84–85, 136–37
Huntington, P. D., 107–8

Iberia, Ohio, 154
Illinois, 66, 114, 203
images, 69–70, 72, 86, 92–96, *97*, 208; of class distinctions, *51*; of conductors, *83*; on currency, *94*, *95*, *96*; of depots, 30, *48*, *202*; of inexperienced travelers, *48*; of Jenny Lind, *36*; in panoramas, 96–97, 207; in political cartoons, 92, *93*, *94*; of railroads, 4, *34*, *36*, *48*, *81*, *87*, *88*, 89–90, 92, *93*, *94*, 95, *96*, *98*, 100, 109, *110*, *191*, *195*, 201, *202*; with religious messages, 122; of school, *191*; on sheet music, *87*, *88*, 89–90; of steamboats, 4, *51*, *93*, *95*, *96*, *98*; with temperance messages, *34*, *110*
Indiana, 137, 146
Indianapolis, Indiana, 43
Iowa City, Iowa, 114
Ithaca, New York, 189

Jerry (an enslaved man), 35
Johnson, Herschel, 92, *94*

Kauffman, Anna, 60
Kendall, Amos, 58
Kent county, Maryland, 153
Kentucky, 46, 52, 57, 84, 135, 145, 147, 154
Keokuk, Iowa, 143
Kirby, Edmond, 41
Knowlton, J. H., 109
Knox County, Illinois, 19
Koren, Elisabeth, 52, 169

Lake Erie, 57, 122
Lake Michigan, 42
Lane, Joseph, *94*
Langdon, William Chauncy, 62
Laurens, Henry, 20
Lewiston, Maine, 89
Lexington (steamboat), 69–70, 124–125
Lexington, Kentucky, 57
liability, 28, 32–33, 63–64, 82
Lieber, Francis, 43

Limaville, Ohio, 154
Lincoln, Abraham, 92, *94*, 135
Lind, Jenny, 11, 35, *36*
Little Miami Railroad, 18
Little Rock, Arkansas, 79
lodging. *See* hotels
Loguen, Jermain Wesley, 162
Long, John, 189
Longgley, John, 193
Long Island, New York, 200
Louisa Railroad, 168
Louisiana, 139–40
Louisville, Kentucky, 152
Lowell, Massachusetts, 13
Ludlow, Massachusetts, 25
luggage. *See* baggage
Lyon, Mary, 59, 173

Mackay, Alexander, 116
Mackenzie River, 194
Mackinac Island, 42
Madison, Dolley, 76
Madison, James, 42
Magnolia (steamboat), 97
Maryland, 62, 139, 152
Mason, R. B., 42
Massachusetts, 167; transportation companies in, 22, 23, 27, 32, 35, 45, 117–18; travel in, 13, 20, 96; travel to, 40
McDermott, Elizabeth, 196–97
McElveen, A. J., 134
McElwain, David, 22–23
McNeill, Sallie, 1–2, 4–5, 8–9, 182, 210
Melville, Herman, 75
Memphis and Charleston Railroad, 16, 63
Memphis, Tennessee, 16, 43–44, 138, 143, 145
Mersman, Joseph, 62, 90, 92
metaphors, 2, 7, 72–77, 102, 108, 187, 198, 208; in literature, 73–75; in private writing, 76; steam as a, 73–74, 149–50, 207
Michigan, 28, 100
Michigan Central Railroad, 28, 50, 171
Millburn, Mary, 145
Minnesota, 20, 203
Mississippi, 3, 144, 194, 199
Mississippi River, 16, 69, 95–96, 114, 138, 203
Mobile, Alabama, 62
Monroe, James, 42

Index

Mooney, Thomas, 178
Moore, Andrew Charles, 32
Mott, Lucretia, 174
Murdock, Rev. Dr., 74
Murray, Rev. N., 116–17
music, 7, 72, 86–92, 102, 208; for commemoration of steam transit, 86; imitative of steam transit, 2, 87, 90–92; lyrics in, 60, 89–90; orchestral, 90, 92
Myers, Jeremiah, 160

Nashua and Lowell Railroad, 25, 27
Nashville, Tennessee, 143–44
Native Americans, 5, 95, 102, 193–94, 203
New Albany and Salem Railroad, 146–47
Newburyport, Massachusetts, 29, 118
Newcastle, Delaware, 143
Newfoundland, Canada, 125–26
New Hampshire, 21, 27, 99, 191, 210
New Jersey, 147, 191
New London County, Connecticut, 34–35
New Orleans, Jackson, and Great Northern Railroad, 89
New Orleans, Louisiana, 46, 62, 87, 89, 91, 136–39, 145
New Orleans and Great Northern Railroad, 87
Newport, Rhode Island, 38
newspapers, 13–14, 54, 64, 113, 173, 192; on depots, 30; on railroads, 91–92. *See also* Black mobility
New York (state), 101, 140, 148, 191; transit through, 20; travel from, 46; travel to, 38
New York and Erie Railroad, 74, 100
New York and Harlem Railroad, 161
New York City, New York, 132; accidents in, 67, 198; dangers to travelers in, 68–69, 171; in fiction, 80, 203, 205; musical performances in, 91; railroads in, 28; travel from, 83, 113, 123, 158–59, 163, 187; travel to, 59, 151, 187; Underground Railroad in, 151–52Niagara and Buffalo Railroad, 160
Niagara Falls, 155, 203, 206
Nichols, Charles W., 127
Norfolk, Virginia, 137, 145
North, Jane Caroline, 55–56
North Berwick, Maine, 74
North Carolina, 12, 27, 123, 136, 140
Northern Railroad, 8, 32, 87

Norwich and Western Railroad, 14
Norwich and Worcester Railroad, 45, 168
Noyce, Elisha, 198

Obear, J., 121
O'Brien, Andrew Leary, 67
Ohio, 18, 60, 100, 101, 122, 148, 153
Ohio River, 133, 138, 144, 146, 148, 210

Pacific Railroad, 106
Parker, Martha, 42, 168
Paxson, Joseph, 190
peddlers, 54–55, 61, 83
Pennsylvania, 100, 148
Pennsylvania Central Railroad, 89
Pennsylvania Railroad, 24, 100
perceptions, of distance, 43, 65; of speed, 42–43, 59, 71, 74, 78, 87; of time, 43–44, 62, 65, 74–76, 182, 184
Petersburg, Greensville and Roanoke Railroad, 27
Philadelphia, Easton, and Water-Gap Railroad, 24
Philadelphia, Pennsylvania, 127, 137, 143, 181, 190, 192; as gateway to freedom, 148; transportation companies in, 18; travel from, 45; travel to, 143, 145, 151, 158, 161
Philadelphia, Wilmington and Baltimore Railroad, 29, 45, 54, 100
Pittsburgh, Pennsylvania, 46, 120–21, 148
Pittsfield, Massachusetts, 33
Plummer, Frederick, 114
poetry, 14, 65, 69, 195, 200. *See also* fiction
politeness. *See* etiquette
Pond, John F., 30
Poole, Caroline, 13
Pope Gregory XVI, 104, 110
Pope Pius IX, 110
Porter, Lemuel, 127
porters, 47–49
Portland, Maine, 62, 190
Portsmouth and Roanoke Railroad, 26
Poughkeepsie, New York, 83
Powers, Hiram, 173
Protestants: and anti-Catholicism, 104, 109–12; reaction to accidents, 39, 104, 117, 123–28; skepticism of steam transit, 104, 116–123; support of steam transit, 103, 105–7, 109–10
Providence, Rhode Island, 47

Providence and Worcester Railroad, 38
Public vs. private space, 165–69, 175, 177–79, 182, 185–86, 209

Racine, Wisconsin, 153
railroad conductors, 82–83, 85; discrimination by, 159–63; and enslaved persons, 135–36, 140–41, 144, 146–47, 150; as a position of responsibility, 49–50, 171–72, 181, 204
railroads: celebration of, 14–16, 34, 86; as common highways, 11, 17; construction of, 12–14, 18–19, 21, 24, 64, 189–90; derailment of, 28; first trips on, 1, 6, 13, 42–43, 187, 208; land acquisition for, 6, 11, 18, 21–24, 26, 190. *See also specific corporations*; steam transit; railroad conductors
Raleigh and Gaston Railroad, 27
Randolph, Evan, 13
Ray, Charles, 158, 163
reading, 1, 54–55, 114–15, 180–81
religion, 103–28, 209. *See also* Protestants
Resseguie, Anna Marie, 125
Rheem, Eli, 192
Rhinebeck, New York, 57
Richmond and Danville Railroad, 16
Richmond, Fredericksburg and Potomac Railroad, 29
Richmond, Virginia, 29, 55, 134, 143
Riddle, Dr., 120
Rix, Chastina, 19, 190
Roberts, Solomon, 17
Rochester, New York, 152
Ruggles, David, 159–60, 163
Russell, Charles, 21

Sabbatarianism, 7, 118, 120–23
Salem, Ohio, 156
Sandusky, Ohio, 147
Sanford, Mollie Dorsey, 63
San Francisco (steamboat), 126
Saratoga Springs, New York, 43
Savannah, Georgia, 34, 147
Saxon, H. M., 114–15
Schenectady and Troy Railroad, 45
school, 190–91
schoolbooks, 192–94, 196, 198–99
Seabury, Carolina, 56
Second Great Awakening, 103, 196

senses, 3, 7, 39, 63, 71; descriptions of, 36, 44, 58–59, 121, 187, 199; disorientation of, 201. *See also individual senses*
Seward, Frederick William, 76
Seymour, Indiana, 143
Shaw, J. B., 127
Sheldon, Asa, 24, 167
Shreve, Thomas, 73
sight, 1–2, 14, 32, 59–60
slavery: abolition of, 149–57; confrontation about, 131–32, 152; confrontation at depots, 8, 146–47, 154–56, 163; confrontation at wharves, 8, 153–54, 163; emancipating enslaved people who traveled north, 155–57; enslavers pursuing enslaved people in the North, 151–55; escaping by hiding aboard steam transit, 145–47; escaping by passing as white, 144–45; escaping by sending self as freight, 143–44; escaping using disguises, 144; escaping via steam transit, 131–49, 163, 209; escaping with employees' assistance, 138, 146–47, 151. *See also* Underground Railroad
Smell, 59, 71
Smith, George, 43
Smith, Sara Chester, 165
Sound. *See* hearing
South Carolina, 95, 109, 135, 140, 155, 172
South Carolina Railroad, 20, 135
Speed, Elizabeth, 189
Springfield, Massachusetts, 25, 191
Stanton, Elizabeth Cady, 74, 174
Stark, George, 25–27
steamboat captains: and discrimination, 157, 159–62; as a position of responsibility, 49–50; as protagonists in fiction, 82–83; reckless, 66–67
steamboats, 3, 42; and competition among towns, 25; and conveyance of freight, 27; first trips on, 6, 13, 41–42. *See also specific steamboats*
steam transit: advertisements for, 12; alternate uses of, 4, 9–11, 32–34, 107–8, 114, 131, 209; animal reactions to, 13, 21, 37; changes to landscape by, 16; and children, 187–207; and Christian morality, 117–20, 202–3; and class distinctions, 50–52, 135–36, 158, 161; and

community engagement, 14–16; discrimination on, 8, 132, 152, 157–63, 174, 209; as an economic tool, 2, 7, 10, 14, 20, 30, 77, 208; and enslaved labor, 8; and enslaved people, 2, 134–64; impact on land values of, 19–20, 30; as modern, 3–4, 183, 201–2; and moral lessons, 8, 188, 194, 196–207, 209; naturalization of, 1, 9, 34–37, 71–73, 102, 128, 164, 182, 186, 207, 209; as a novelty, 1, 10–12, 41–43, 190, 196–97, 203–204; as a political tool, 2, 10, 34, 208; projected benefits of, 6, 13, 195–96, 208; and risk of idolatry, 119–20, 123, 126; as a sign of civilization, 3, 95, 101–2, 105, 107, 193–94; as a sign of God's favor, 7, 103–12, 128; skepticism or opposition to, 16, 18–19; and the slave trade, 3, 7, 135–136; used to spread religion, 7, 106–8, 112–16, 121, 128; towns' restrictions on, 29–30; value of, 188. *See also* children; images; metaphors; newspapers; railroad conductors; railroads; senses; slavery; steamboat captains; steamboats; talking to strangers; travel; women
Steyermärkische Gesellschaft, 90–91
St. Louis, Missouri, 60, 137, 142
Stone, John, 125
Stonington Railroad, 160
Stowe, Harriet Beecher, 75
Strong, George Templeton, 35–36, 44, 76, 178
Susquehanna River, 45
Sykesville, Maryland, 62
Syracuse, New York, 35, 156

talking to strangers, 55, 71, 174; about political topics, 56; about religion, 115–16; about slavery, 56–58
Tappan, H. P., 111
taste, 40–41, 61–62, 74, 87, 177, 198
Taylor, Zachary, 190
Telfair, Mary, 62
temperance, 33, 75–76, 109, 113. *See also* images: with temperance messages
Tennessee, 134
Texas, 1
thieves, 49, 68–69
Thompson, Richard, 43
Thoreau, Henry David, 16
Thorne, Samuel, 190

time: management, 39, 45–46, 65, 180–81; regulation of, 46, 108, 190; spent, 54, 56; value of, 197
Titus, Charles, 65
Toronto, Canada, 148, 203
touch, 62–63
toys, 190, 210
travel: advice, 179–85; commodification of, 4–5, 8, 38, 44–47, 71, 99, 208; guidebooks, 7, 62, 64, 68–70, 72, 74, 97, 99–102, 209; by horse, 6, 14, 40–41, 50, 63; intermodal, 5, 13, 38, 44–46, 95; by stagecoach, 13, 41–43, 50, 63, 157, 165, 184; by walking, 6, 13, 31–33, 50, 63. *See also* Black mobility
travelers: experienced, 49, 68–69, 187, 206; inexperienced, 48, 49, 68, 87, 179–183
Turpin, Walter Gwynn, 20
Tyler, John, 154

Underground Railroad, 2, 8, 86, 89, 132–33, 145, 148–50; in music, 89–90. *See also* slavery
Utica, New York, 45, 47

van Bibber, Isaac, 62
Vanderbilt, Cornelius, 124
Vermont, 19, 21, 84, 191
Verplank, Jason, 152
Virginia, 20, 140, 155; travel companies in, 26–27, 168; travel from, 43, 136
Virginia and Tennessee Railroad, 16, 20–22, 141

Waller, Henry, 42, 44, 49, 62
Waller, Sarah, 173–174
warning signs for trains, 32, *195*. *See also* danger; hearing
Warren, Massachusetts, 191
Washington, DC, 13, 43, 91, 144
weather, 33, 40, 62–63, 84
Webb, Augustus Pleasants, 41, 54
Western and Atlantic Railroad, 20
Western Railroad, 18, 33, 49, 117–19, 168; construction of, 13, 25; employees or, 127; and land acquisition battles of, 22–23
wharves, 68, 102, 170, 197; alternate uses of, 35. *See also* slavery: confrontation at wharves
Whately, Massachusetts, 30

Wheeling, Virginia, 87
White, Caroline Barrett, 54, 59, 68, 126, 174
Whittier, John Greenleaf, 57
Wilbraham, Massachusetts, 25
Williams, William, 40
Wilmington, Delaware, 45
Wilson, Joseph, 106
Wilson, William Hasell, 18
Winchester, Virginia, 46
Wisconsin, 114, 143, 187
Wolff, Joseph, 113
women, 8, 209; and chaperones, 8, 170–73, 175, 185; and new opportunities found in steam transit, 167–69, 171–72, 174, 181–83, 185; and separate spaces on steam transit, 167–70, 186; white, 165–86
Worcester, Massachusetts, 14, 30, 38, 90–91, 168, 174
Wright, Dexter Russel, 56, 76
Wright, Elizabeth Steele, 13, 40–41

Xenia, Ohio, 154

York, Pennsylvania, 26, 192
Young, Anna Rebecca, 77
Young, Thomas John, 56

Zanesville, Ohio, 154

Milton Keynes UK
Ingram Content Group UK Ltd.
UKHW031321190924
448427UK00002B/24